Advanced Quality Function Deployment

Advanced Quality Function Deployment

Fiorenzo Franceschini
Professor of Quality Management
Department of Manufacturing Systems and Economics
Turin Polytechnic
Turin, Italy

CRC Press
Taylor & Francis Group
Boca Raton London New York

CRC Press is an imprint of the
Taylor & Francis Group, an **informa** business

CRC Press
Taylor & Francis Group
6000 Broken Sound Parkway NW, Suite 300
Boca Raton, FL 33487-2742

ISBN 13: 9781574443219 (hbk)

Visit the Taylor & Francis Web site at
http://www.taylorandfrancis.com

and the CRC Press Web site at
http://www.crcpress.com

Library of Congress Cataloging-in-Publication Data

Franceschini, Fiorenzo.
Advanced quality function deployment / Fiorenzo Franceschini.
p. cm.
Includes bibliographical references and index.
ISBN 1-57444-321-6 (alk. paper)
1. Quality control. 2. Production management—Quality control. I. Title.

TS156 .F73 2001
658.5'62—dc21 2001048518
CIP

Dedication

To Anna Maria, my wife,
and to Piero and Giorgio, my sons,
for their continuous support

Dedication

Foreword

The quality of a product or a service, understood as its capacity to meet customer needs, stems from and gains substance even in the initial stages of project planning. This concept has resulted in applied research channeling of numerous efforts toward the implementation of new tools aiding the activity of design.

This text, in line with these assertions, intends to present and discuss one of these tools, quality function deployment (QFD).

In this work the basic ideas underlying the methodology are described, as well as the innovations introduced and the elements of stimulus brought to the new science of design.

My particular thanks for the realization of this work go to Professors Sergio Rossetto, Raffaello Levi, and Anthos Bray; and Doctors Marco Terzago and Maurizio Galetto.

Preface

A preface may have a twofold purpose, namely, to condition and to clarify. In effect, the authority of the person attesting the validity of the work and the capability of the author somehow affect the *a priori* judgment of the reader. Furthermore, the same preface is aimed at bringing out the motivation underlying the author's effort. Given these premises, it is quite clear why presentation is an absolute must for a poor text, and why, on the other hand, a good book can fare quite well without any viaticum.

Because Professor Franceschini's work definitely belongs to the latter category, this preface can easily be dispensed with. The reader may find out that the work is original in both layout and contents just by browsing over the text. Then, going over but a few pages, the reader will be pleased to discover that a technical subject can be dealt with in a clear, albeit rigorous, manner.

However, I am glad to write these few lines because by doing so I can testify that quality has been brought back squarely where it belongs, namely management engineering, the one and only science entitled to treat it as part and parcel of systemic company management. Given that quality and innovation are in many ways synonymous, systematic and dynamic valence of quality then follows. Furthermore, there is no doubt that quality has a complex and dynamic dimension, requiring for its management the harmonious and determined concourse of the entire company. Complexity stems from encompassing a *multiplicity of dimensions* (expected quality, offered quality, perceived quality) as well as a *multiplicity of attributes*.

Its dynamism stems from the fickleness of market expectations and from the pressing game competitors are wont to play, these among the main premises of Professor Franceschini's work. Starting with the analysis of techniques best suited to evaluate and link the customer's needs to the technical characteristics of a product, he turns the focus to quality in services, showing clearly how awkward an objective evaluation of attributes may be (seldom allowing for objective measurements) and showing effective evaluation and exploitation methods.

The subject is extremely interesting on a purely speculative level as fits a current research topic, and on a practical level too, because it concerns, over and above the tertiary sector, also manufacturing firms, which to market their products must provide their customers with a comprehensive series of collateral services that combine to form the overall quality of the product sold.

I conclude by wishing the work the success it definitely deserves and the reader a fruitful reading.

Sergio Rossetto*

* Vice-Chancellor of the Polytechnic Institute of Turin; Director of the Polytechnic School of Economics and Organization "Vilfredo Pareto".

About the Author

Dr. Fiorenzo Franceschini is professor of quality management at the Polytechnic Institute of Turin — Department of Manufacturing Systems and Economics. He is author and co-author of three books and numerous papers presented in scientific journals and at international congresses. He is a member of the Editorial Board of *Quality Engineering* and *International Journal of Quality and Reliability Management* journals. His current research interests are in the area of quality engineering and control, quality function deployment (QFD), service quality management, and industrial metrology. He is a senior member of American Society for Quality (ASQ) and the Institute for Operations Research and Management Sciences (INFORMS); and a faculty member of Consortium of Universities in Quality Engineering (QUALITAL). Since August 1997, he has been a member of the European Experts Database as evaluator of the research technological development (RTD) proposals in industrial and materials technologies for the European Community.

Table of Contents

Chapter 4

Chapter 5

Chapter 10

Chapter 11

Chapter 12

1 Quality and Innovation — Conceptual Model of Their Interaction

1.1 INTRODUCTION

For many years, Japanese industry has been analyzed to investigate the reasons for its success and to evaluate the transferability and adaptability of the model to the Western world. This model has been studied by engineers, economists, and sociologists, each giving a particular contribution to the understanding of the phenomenon.

Some have identified the reason for Japan's success as making better use of technological innovations (its own or others), including the automation of manufacturing processes. Others have identified the strength of the Rising Sun as its greater confidence in the future, confidence to favor long-term investments, but with a high innovation rate.

Engineers and economists [Abernathy, 1971] together have asserted that Japanese success is based on huge investments in research and development (R&D) and a marked rapid application of the obtained results. Politicians and industrialists [Dore, 1991] have repeated that the better fortune must be attributed to a rather "rogue" commercial attitude: exporting with below-cost prices and imposing obstacles of any type to imports. Sociologists [Mills, 1954] have hypothesized that ethnic uniformity, social peace, and high-level education are at the basis of the long Japanese spring. Finally, entrepreneurs have identified the main causes of their success as massive public support and low manpower cost.

Today, thanks to the better and widespread knowledge of the Japanese situation, these judgments have receded a little. Thus, we recognize, for example, that Japan does not generally possess the automation level of Western factories and that Japanese entrepreneurs do not refrain *a priori* from pursuing short-term policies.

At the same time, the idea that their success must be a result of government grants and low manpower cost has lost ground. The former is not greater and the latter is not less than those that may be found in the other industrialized countries. An analogous reduction has suffered from the read capability of the market: the Japanese, like their competitors, have no particular analytical or anticipatory endowment. Years ago they selected some market sectors in which to operate, those with lower investments and higher repayments; in these sectors, they have tried to acquire a monopolistic position, to "drive" and not to "suffer" the market, and hence to start their economic takeoff.

After a long debate, many agree that the Japanese success is based on the binomial *quality–innovation*, where quality is a multiattribute function [Garvin,

1987; Huthwaite, 1988, Franceschini and Rossetto, 1995; Galetto, 1996] involving any element that makes a product more desirable for the customer; and innovation is recognized as any intervention that can modify, even if only marginally, the market [Villa et al., 1991].

Both quality and innovation are looked on as dynamic functions coupled in a continuous evolution. The assumption of a strong correlation between these two functions immediately lays the basis for two distinct problems. The first one is of a theoretical nature and the second, of a practical nature. The theoretical one concerns the construction of an explicative conceptual model able to correlate quality and innovation and to explain their dynamic nature. The other concerns the way in which an enterprise can execute the model.

1.2 QUALITY AND INNOVATION

It is evident that not the innovative intervention, but its effects on a good attract the customer. Any action able to augment customer judgment of the offered good is concretely innovative: not only those increasing product performance but also those improving, for example, delivery time, after-sale service, or product image.

These effects, perceived and evaluated in an ordinal or cardinal manner, are transformed by the customer in a set of attributes that together define the *perceived quality* (Q_p) of a good. The customer judges and chooses a product on the grounds of its quality, which therefore is the main cause of its commercial success.

In a nearly axiomatic form, it follows that the effect of the innovation is the improvement of quality, which itself becomes the aim of innovation [Villa et al., 1991; Franceschini and Rossetto, 1995].

Even though what has been said couples quality and innovation, it still does not explain why the two functions are dynamic. To understand this, the quality concept must be analyzed in more detail.

In fact, in addition to the perceived quality there is the quality that is actually assured by the producer through its design–manufacturing–commercial system. The latter is the so-called *offered quality* (Q_o). Generally the two qualities Q_p and Q_o are not equal, because of the information asymmetries and the different metric used to evaluate the product attributes. Customer evaluation is based on a reference model that compares different products, and is subject to the marketing pressure of all competitors. Generally, this model leads to the so-called *expected quality* (Q_a), which for its changeable nature does not coincide with the Q_p.

To preserve and to increase its own market share, every enterprise must direct its effort to modify all three dimensions of quality (Q_a, Q_o, Q_p), in such a way that Q_o approaches both Q_p and Q_a. To achieve this goal an enterprise must develop innovative interventions. On the other hand, because every enterprise has the same problem, all behave alike; thus, Q_a and Q_p, as effects of the free market competition dynamics, appear as time variable functions. As a consequence, innovation cannot be an isolated action, but is a continuous dynamic process.

A first schematic representation of the innovation process is as follows. An enterprise evaluating the difference ΔQ between Q_a and Q_p develops two complex actions to increase customer judgment of its product. The first, through

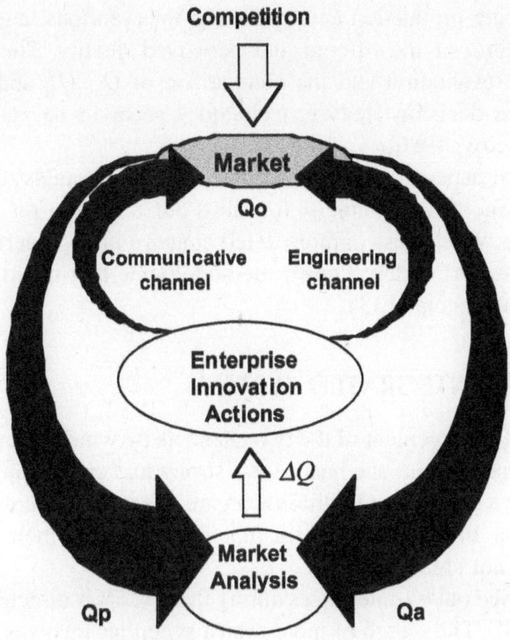

FIGURE 1.1 Schematic representation of the innovative process.

marketing, is directed to conditioning the customer; the second, through a series of technical–organizational interventions, is focused on improving the designing–manufacturing–supporting system so as to obtain an intrinsically superior product. The two actions are always present, although the relative intensity depends on the market sector, the maturity of the product, and the customers' culture. In any case, to intensify the results both actions must be coordinated by means of an adequate systemic approach in enterprise management.

From a control-science point of view, the innovation process can be synthesized in two distinct feedback loops. The first one has a prevalent communicative–persuasive content, and the second, a prevalent engineering–organizational character (Figure 1.1).

The communicative–persuasive channel, managed by the marketing function, has the target of modifying Q_p and of inserting in Q_a some peculiar attributes of the offered product. The main aim of the engineering–organizational channel, on the contrary, is to improve Q_o.

If the conceptual model has some appeal for its theoretical use, two main problems have to be solved. The first one concerns the construction and the identification of:

$$\Im\left(\dot{Q}_o, Q_o, PI\right) = 0$$

and

$$\aleph\left(\dot{Q}_p, Q_p, MI\right) = 0$$

where *PI* and *MI* are production and marketing interventions, and $\dot{Q}_o$, $\dot{Q}_p$, the first derivatives over time of the offered and perceived quality. The second problem concerns both the evaluation and the comparison of Q_a, Q_p, and Q_o. For the last point, multicriteria decision analysis techniques seem to be particularly suitable [Ostanello, 1985; Roy, 1996].

The theoretical aspects aside, in the next section we analyze how enterprises activate the two operative channels, focusing our attention on the engineering–organizational one. We discuss in more detail concurrent engineering (CE) and lean production (LP) as two widely known methodologies that together lead toward a lean and integrated system (LIS).

1.3 LEAN AND INTEGRATED SYSTEM

The activation and management of the two channels by which an enterprise interacts with the market asks, on the one hand, for a strong and coordinated intervention. It requires an adaptive agility that traditional organizational structures, with their clean and stiff separation among the different functions and with their very long hierarchical chains, are not able to assure.

The effort carried out by enterprises during these years is directed to transforming themselves into LIS. The search of more robust synergies involves a revolution both in internal structure and in external relations. Inside, such a revolution is realized with the progressive dismantling of bureaucratic structures stratified by time and size of the enterprise, and of the nonessential hierarchical levels that penalize the decision process. Outside, the revolution brings new relationships both with suppliers, no longer seen as servers but as partners involved in the enterprise strategies, and with customers whose satisfaction becomes the primary target.

Lean and integrated are the enterprises in which the only functions present are those that add value, in which wide horizontal decision spaces are available; process visibility is complete, and friendly and cooperative relations exist, not only among the different internal functions but also with customers and suppliers.

The progressive approach to customers allows the enterprise to take into consideration their judgment during the designing or redesigning phases of a product, resulting in a better approximation of Q_o to Q_a. Moreover, because suppliers are directly involved with the objective of the enterprise, LIS presents less distinct borders than a traditional one, and its management cannot be based on a hierarchical structure of a classical and rigidly sectored type. The peculiarity of LIS is to recognize, as enterprise encrustations, organizational elements that were believed necessary; and to propose, in an industrial and modern way, the *Bottega Rinascimentale* (Renaissance Workshop), which gathers all the necessary skills to execute their work giving due attention to the voice of the customer.

Unfortunately, there are many organizational and cultural obstacles to the practical application of LIS concepts: from the single specialist to the multispecialized groups culture, from hierarchical organization to management by objectives or processes and so on (a century of industrial history constitutes an inertial mass that is difficult to move). A finger is pointed toward managerial style [Feigenbaum, 1991;

Maslow, 1954; Mills, 1954], which has been subjected to severe criticism concerning essentially: the opportunism at the basis of the internal relations, the motivation always connected to tangible forms of remuneration, and the so-called role of regulation that penalizes management by objectives strategies.

Because of the previously mentioned difficulties, if the enterprise had not rediscovered the centrality of producing real goods, downplayed in the recent past by advantaging both financial and purely commercial activities, this new model would have not attracted the attention of the Western world. In fact, general guidelines proposed in the LIS have found their natural application field in the manufacturing environment, to which CE refers to the designing phase and LP refers to the production cycle.

1.3.1 Concurrent Engineering

For a long time, many enterprises, taking advantage of the tendency of the market to be stationary and have low turbulence, have operated mostly with the communicative–persuasive channel, in such a way as to assure a satisfactory reception by the customers and then a longer life to their own products. However, if the communicative lever is always important, especially in the low technology mass products sector, it is true that by itself it does not adequately guarantee competitiveness in those sectors where the technological component is not negligible and where the competition shows great aggressiveness.

In these markets, the possibility of an enterprise achieving and maintaining a lasting leadership is tied both with the capability of offering a real quality Q_o and the ability of renewing the products at a fast rate. CE enterprises understood the need, before assuming in 1989 the present denomination [Abernathy, 1971; Hartley and Okamoto, 1998], to give some answers for the design phase of these questions.

Western enterprises progressively understood design to be the main element responsible for quality Q_o, and for costs supported throughout the life and time to market of the product. Concerning costs, it is important to remember that the design phase, although contributing on average only 5% of the total product cost, is responsible for about 75% of the overall manufacturing cost, for about 70% of its life cycle cost and for over 80% of its qualitative characteristics [Huthwaite, 1988; Nevins and Whitney, 1989; Dowlatshahi, 1992].

Concerning the time to market, design contraction offers some important competitive advantages:

- On the one hand, a shorter time allows lower investments and, therefore, asks for a shorter payback period with a reduction in risk.
- On the other hand, a shorter time to market allows one to drug the market, artificially accelerating competitors' product aging and then damaging them under the commercial profile.

The possibility of CE improving the design phase of a product, depends on its initial consideration not only of its primary functions but also of its aesthetics, producibility, assemblability, maintainability, and recyclability.

A definition of CE is given by the Institute for Defense Analysis in report R-338:

> Concurrent Engineering is a systematic approach to the integrated, concurrent design of products and their related processes, including manufacture and support. This approach is intended to cause developers, from the outset, to consider all elements of the product life cycle from conception through disposal, including quality, cost, schedule, and user requirements.

It follows that CE is a new organizational and managerial approach in which all professional skills that support the product during its life cycle are activated, so as to transform customer *desiderata* (desire) in product specifications [Sohlenius, 1992; Sweeney, 1992]. CE, therefore, avoids triggering a traditional and penalizing serial and iterative product design path resulting from the clear-cut division of jobs, as in the traditional phase-review process [Kusiak, 1993]. However, for the correct introduction of CE into an enterprise many difficulties must be overcome. Some of these are related to managerial style as we have said; others, on the contrary, are linked to technical issues.

Managerial style is the cause of difficulties in jointly involving different competencies in the same design process. Technical difficulties depend on the heterogeneity and the complexity of information that must be gathered, managed, stored, retrieved, updated, and decoded for a real and effective integration. Such difficulties sometimes force exhaustive and often inconclusive meetings, where all problems are again taken from the beginning and where participants intervene in a very approximate manner on the basis of a hurried verbal updating.

To facilitate parallelism to the design in both a product and its manufacturing processes, some hardware and software tools and various methodologies (quality function deployment [QFD], relational databases, elaborate procedures for documentation management, and so on) have been developed.

QFD [Akao, 1986] guarantees that customers' requirements are heeded from the beginning, when the decisions concerning product characteristics are made. This methodology also provides the comparison of these characteristics with those of other products (benchmarking process) [Zairi, 1992], in such a way that the desired competitive level can be established *a priori*.

Another tool is *design for assembly and disassembly*, which is able to support designers' efforts to reduce product complexity without compromising its functionality [Boothroyd, Dewhurst, and Knight, 1994]. Other techniques are *manufacturing and assembly* and *manufacturing capability deployment*; the former helps designers in the analysis of particularly complex projects, and the latter facilitates manufacturing system selection [Sweeney, 1992]. Also *rapid prototyping* [Kruth, 1991; Jacobs, 1992] has been added to the list of the available techniques for CE. This technique allows the production of an artifact straight from digital models built with computer-aided design tools. The resulting plastic objects are useful during the initial phases of the design process, and they find a useful application in the prototyping and testing phases, if the material they are made of corresponds to final component specifications.

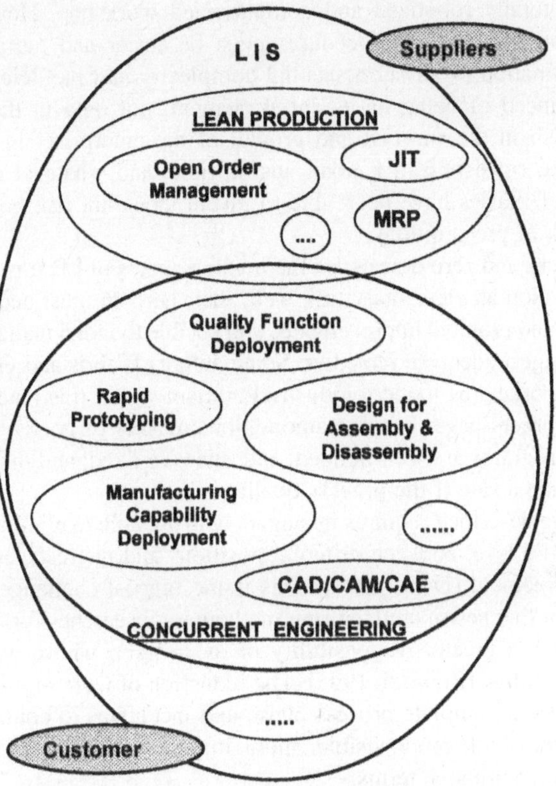

FIGURE 1.2 Lean production and concurrent engineering environments.

Although CE is effective for both quality improvement and time-to-market reduction, it does not complete the engineering–organizational channel. In fact, CE is an integrated methodology to design a new product and its manufacturing cycle, but it is inadequate to grasp and to eliminate all the inconveniences that arise during the manufacturing phase. This is a task of the manufacturing function. The methodology that has taken shape during the 1980s to handle this is LP, which together with CE constitutes LIS (Figure 1.2).

1.3.2 Lean Production

If the main aim of CE is the easing of a product design, LP must guarantee that a product is built in the correct way at the estimated cost. It must also facilitate the improvements supported by a daily practice lived in a participative manner. For LP to be realized, all LIS guidelines must have practical application. First of all, the worker must have an active and positive part in the manufacturing process.

This is certainly the first and perhaps the most important undertaking at the basis of lean production [Womack et al., 1991], which by means of a renewed anthropocentric vision of the factory (*human integrated manufacturing*) puts an end to the dream of the totally robotized and computerized workshop. However, for better involvement of workers, all procedures must be easier and more visible, with a complete elimination of any nonessential complexity and inefficiency. The worker must be convinced of being an essential element, not only in the manufacturing process but also in the survival and growth of the enterprise. In such a way the Japanese menu, consisting of kanban, just-in-time, and whatever else consultants' and imitators' fantasies have been able to invent, does not risk becoming a sterile operating book of prescriptions.

If zero stocks and zero defects are the mythical goals of LP, it is not sufficient for workers to develop an elementary task well; their horizon must become wider; they must maintain and promote improvements, and be able to individualize flaws, pinpoint causes, and suggest adequate remedies. Some authors [Fabris and Garbellano, 1993] look at these workers as modern industrial craftsmen, but this reading is excessive because the workers never become autonomous makers of goods: production time is measured, programs are well-defined, and rules are fixed; and the rule itself is the freedom to elude a rule if the process quality asks for it.

All we have described requires managers who are able to eliminate conflicts and stresses, to find a level between different positions and interests, and to generate a widespread agreement [Dore, 1991]. This is the biggest challenge of LP, because without a doubt this new manufacturing methodology searches for an extraordinary commitment and a greater responsibility of the worker, whose work comes back into the shop window [Bonazzi, 1993]. The reduction of *work in process* (WIP), the responsibility for a complete process phase and incentives to continuous improvement, makes the work more visible, more inspectable, and more measurable in qualitative and quantitative terms.

LP imposes radical mutation not only of the worker's role but also of intermediate cadres, who are pushed toward being less bureaucrats and more managers, able to provide solution to problems.

This requires a considerable cultural renewal for all people operating in a factory: the worker must relinquish the merely executing role; the intermediate level, the bureaucratic practice; and the top management, the hierarchical stiffness and the excess of abstraction.

Although we have spent a long time in framing the worker's role, because this is the most innovative issue of LP, it is important not to forget the role of suppliers. They must become an integral part of the enterprise. Such a result can be obtained by involving suppliers from the design phases of a product, so as to give wide knowledge about the needs and objectives of the enterprise with which they are partners and to develop adequate contracts, such as to reward quality and punctuality.

To give consistency to this multiple involvement many tools have been conceived: from *brainstorming* and *brain-writing* tools to the *group decision support system* (GDSS), and so on [Fabris and Garbellano, 1993]. It is important to underline that to achieve the desired targets each enterprise must trace a specific path, taking into account its own current status, peculiar recent history, and conditions of its operating environment.

In any case, starting LP is possible only if there is convincing and lasting acceptance by all participants, and not a passive translation of Japanese rules, for which it probably will follow that "not all that was new was also good and not all that was good was really new" [Bisgaard, 1989].

1.4 CONCLUSION

Enterprises engage in a continuous process of innovation because they need to offer competitive quality. The more an enterprise approximates to LIS the better its results are. In fact the desired efficiency for the communicative–persuasive and engineering–organizational channels can be assured with slim structures and coordinated activities, as well as by better relations with customers and suppliers.

For better customer satisfaction, enterprises have taken many initiatives, from the creation of marketing information systems oriented to the evaluation of customer expectations to the predisposition of after-sales service networks to solve the customer's problems.

Suppliers have long been considered as passive servers; only recently have they been called both to collaborate with their specific experiences in the development of a new project, and to participate actively in production flow and product improvement. To obtain this new relationship with suppliers, enterprises have introduced new methodologies for their selection and monitoring, and for rewarding their contributions to design and manufacturing problem solutions.

Although CE and LP seem to be the most adequate methodologies for ameliorating design and manufacturing, there are many difficulties to be overcome. With reference to CE it is necessary to overcome resistance to a stronger collaboration between different functions and to find methodologies able to evaluate objectively the contributions of the new design support [Hestand, 1991; Newall and Dale, 1991].

In reference to the production phase, workers must be persuaded of the vital importance of their collaborative presence in the factory, and rewarded if their response is positive. Enterprises are studying different ways of evaluating workers' contributions to the improvement of the product and of its manufacturing cycle, and of rewarding care and results. One possibility is to correlate a part of salary to results [Weiss, 1990], expressed, for example, in terms of productivity and reduction of discards.

However, the problems are much deeper. Our idea is that all the difficulties of LIS implementation are particular aspects of a more general difficulty, which concerns the conception of a new social contract and perhaps of a new way to conceive the capitalist system.

REFERENCES

Abernathy, W.J. (1971), Some issues concerning the effectiveness of parallel strategies in R&D projects, *IEEE Trans. Eng. Manage.*, EM-18(3), 3.

Akao, Y. (1986), *Quality Function Deployment*, Productivity Press, Cambridge, MA.

Bisgaard, S. (1989), Review of Taguchi, 1987, *Technometrics*, 31(2), 257–260.

Bonazzi, G. (1993), *Il tubo di cristallo*, Il Mulino, Bologna.

Boothroyd G., Dewhurst P., and Knight, W. (1994), *Product Design for Manufacture and Assembly*, Marcel Dekker, New York.

Dore, R. (1991), *Bisogna prendere il Giappone sul serio*, Il Mulino, Bologna.

Dowlatshahi, S. (1992), Product design in a concurrent engineering environment: an optimization approach, *Int. J. Prod. Res.*, 30(8), 1803–1818.

Fabris, A. and Garbellano, S. (1993), *Modelli manageriali emergenti*, ISEDI, Torino.

Feigenbaum, A.V. (1991), 3rd ed., *Total Quality Control*, McGraw-Hill, New York.

Franceschini, F. and Rossetto S. (1995), Quality and Innovation: a conceptual model of their interaction, *Total Quality Management*, 6(3), 221–229.

Galetto, F. (1996), *Qualità: alcuni metodi statistici da manager*, Cusl, Torino.

Garvin, D.A. (1987), Competing on the eight dimensions of quality, *Harv. Bus. Rev.*, 65(6), 101–109.

Hartley, J.R. and Okamoto, S. (1998), *Concurrent Engineering: Shortening Lead Times, Raising Quality and Lowering Costs*, Productivity Press, New York.

Hestand, R. (1991), Measuring the level of service quality, *Qual. Prog.*, 24(9), 55–59.

Huthwaite, B. (1988), Designing in quality, *Quality*, 27(11), 111–117.

Jacobs, P.F. (1992), *Rapid Prototyping & Manufacturing*, Society of Manufacturing Engineers, Dearborn, MI.

Kruth, J.P. (1991), Material incress manufacturing by rapid prototyping techniques, *CIRP Ann.*, 40(2), 603–614.

Kusiak, A., Ed. (1993), *Concurrent Engineering*, John Wiley & Sons, New York.

Maslow, A.H. (1954), *Motivation and Personality*, Harper & Row, New York.

Mills, T.M. (1954), The coalition pattern in three person groups, *Am. Sociol. Rev.*, 19, 27–34.

Nevins, J.L. and Whitney, D.E. (1989), *Concurrent Design of Products and Processes*, McGraw-Hill, New York.

Newall, D. and Dale, B.G. (1991), Measuring quality improvement: a management critique, *Total Qual. Manage.*, 2(3), 255–267.

Ostanello, A. (1985), Outranking methods, in *Multiple Criteria Decision Methods and Application*, Fandel G. and Spronk J., Eds., Springer-Verlag, Berlin, pp. 41–60.

Roy, B. (1996), *Multicriteria Methodology for Decision Aiding*, Kluwer Academic, Dordrecht.

Sohlenius, G. (1992), Concurrent engineering, *CIRP Ann.*, 41(2), 645–655.

Sweeney, M. (1992), How to Perform Simultaneous Process Engineering, *Integrated Manuf. Syst.*, 3(2), 15–19.

Villa, A. et al. (1991), Methodological approach to planning and justifying technological innovation in manufacturing, *Computer-Integrated Manuf. Syst.*, 4(4), 114–123.

Weiss, A. (1990), *Efficiency Wages*, Princeton University Press, Princeton, NJ.

Womack, J.P., Jones, D.T., and Roos D. (1991), *The Machine that Changed the World*, HarperCollins, New York.

Zairi, M. (1992), The art of benchmarking: using customer feedback to establish a performance gap, *Total Qual. Manage.*, 3(2), 177–188.

2 Tools and Supporting Techniques for Design Quality

2.1 INTRODUCTION

In recent years quality has shifted from a sectorial goal to a rule of manufacturing life [Garvin, 1987; Franceschini and Rossetto, 1995a]. At the same time, the general attention of enterprises has been progressively focalized on methods and techniques for supporting design [Mattana, 1994; Boschi et al., 1995; Ertas and Jones, 1996].

If on the one hand the thriving growth of new methodologies is a tangible clue of the huge attention directed to the product design, on the other hand, it emphasizes the need of a new conceptual systematization. The choice of how, when, and mostly what to use as a support tool for a product development is still a problem. So too is the ability to evaluate its performances to simplify, speed up, and improve the design cycle.

This issue is not new, of course. An important attempt was made by Pahl and Beitz (1996) to define a sort of designers' reference guide. However, the scenario is rapidly changing with the extraordinary rise of new design methodologies under the systematic stimulus of information technology (IT).

In this chapter we intend to offer a new reasoned and, as far possible, up-to-date survey of tools and supporting techniques for design quality.

2.2 DESIGN AND SUPPORTING TOOLS

Design is a complex and expensive task that, in general, involves both internal company functions (from marketing to manufacturing) and external resources (from consultants to suppliers). Although in the past it was improperly considered as an art, nowadays it has acquired an industrial dimension. It follows the rules of an organized system and it is able to face competitive and selective markets.

The need to reduce the *time to market*, to avoid superfluous costs without affecting the quality, imposes a design process evaluation under two distinct points of view: *technological* and *economic–organizational*. Consequently, these two are the dimensions about which design-supporting tools may find a proper classification and an adequate validation. Besides, if the attempt did not contextually try to correlate these methodologies to specific design activities, it would not get the wanted results.

TABLE 2.1
Design Process Activities and Related Descriptions

No.	Activity	Description
A1	Analysis of market needs and product features	Evaluation of market expectations, definition of preliminary product features
A2	Product functional analysis	Detailed report of product functions and features
A3	Explication of internal and external design activities	Definition of design planning activities, suppliers' role, design criteria and responsibilities, supporting documents
A4	Preliminary design	Feasibility verification according to design specifications and producibility test requirements
A5	Optimization of design parameters, design validation	Evaluation of design alternatives, technical parameters optimization, design validation
A6	Production planning and manufacturing analysis	Technical–economic evaluation of manufacturing process
A7	Design review	Elimination of possible causes for manufacturing and marketing problems
A8	Detailed design	Single parts design and documentation
A9	Product/process engineering	Manufacturing process standardization and simplification, reduction of the number of parts and components
A10	Design qualification	Prototype manufacturing, results verification
A11	Design changes management	Design changes management and documentation updating

By examining literature and empirical case studies [Boschi et al., 1995], we observe that common tasks to all design processes are the spreading use of computer supporting tools and the increasing ascent toward the international reference quality standards [ISO 9000, 1994].

Table 2.1 shows a generic list of the main activities of a design process. This table indicates the order in which the generic life cycle phases of a typical product are started. Each phase is often not totally completed before the next phase begins and several phases may be under way simultaneously [Aurand, Roberts, and Shunk, 1998; ISO 10005, 1996]. According to concurrent engineering (CE) philosophy it is normal to have many parallel or iterative activities. It must also be underlined that some activities may not be present or may have meanings slightly different from those illustrated in the "description" column.

The next step is the description of the most common tools and methodologies able to support design process activities. This is a complex task for many reasons.

First, the high frequency with which new tools are proposed makes obsolete, even at their inception, any attempt at an exhaustive enumeration. Second, a distressing habit of renaming classical tools or techniques makes it hard to appreciate the real news items among the repainted old ones. Third, it is difficult to distinguish between simple academic proposals or prototype tools, and what is an effective new and tested methodology.

A short description of techniques and supporting tools grouped into specific *macroareas*, corresponding to well-defined steps of the design process, follows. Within any macroarea, we define some specific classes to offer a sufficiently clear reference framework. For each tool a brief presentation and some bibliographical references for further investigations are also given.

2.2.1 First Macroarea

The first macroarea is about new design start-up and refers to *market studies* and *quality function deployment* (QFD). Marketing studies concern data analysis methodologies for the estimation of actual and potential market dimensions and sharing among competitors. Well-known methods include *forecasting analysis*, *market segmentation*, *benchmarking* [Zairi, 1992; Bemowski, 1991], and *product briefing* techniques. Data are collected by means of interviews, questionnaires, and comparisons with the competition and marketing channels.

QFD is a functional planning tool used to ensure that the voice of the customer is deployed throughout the product planning and design steps. It represents an adequate environment to carry out a comparative analysis of the technical performance of product with those of market competitors [Akao, 1992; ASI, 1987; Franceschini and Rossetto, 1995b, 1995c, 1997; Franceschini, 1998; Hauser and Clausing, 1988; Wasserman, 1993].

2.2.2 Second Macroarea

The second macroarea concerns design activities that focus their attention toward the economic evaluation, the organization, and the management of a process design. Five classes of these tools are described next.

The first, the *function analysis* class [Pugh, 1991], helps the designer to carefully attribute product functions to each component or subsystem. In this class we find *function analysis and system technique* (FAST) and *function family tree* (FFT) techniques.

The second, the *costs benefits analysis* class, involves the *value analysis* [Miles, 1992] about the problem of superfluous costs reduction, and *value maps* [Urban and Hauser, 1993] for making identification of relationships easier between price and benefits coming from the product usage.

In the same class are *functional cost analysis* [Michaels and Wood, 1989], *economical investments analysis* [Brealey and Myers, 1996], and *risk reduction analysis* [Kahneman and Lovallo, 1993]. The first allows the highlighting of sunk product costs and related causes by means of *activity transaction-based methods* [Ettlie and Stoll, 1990]. The second permits a comparison between costs and incomes of an investment using *discounted cash flow*, *break-even analysis*, and *option evaluation* techniques. The third allows an evaluation of economic risks associated with particular design choices.

The third class involves techniques for planning and project scheduling, and includes *project management* [Kusiak and Belhe, 1993] with PERT/CPM, *work breakdown structure* (WBS), and *flow diagrams* [Brassard, 1989].

The fourth class includes *creative group methods* that are introduced to stimulate the generation of new design ideas, or to solve some specific problems. Typical examples are *brainstorming* and *free* and *forced association* techniques [Hollinger, 1970; Pahl and Beitz, 1996].

The fifth, the *problem-solving* class, involves artificial intelligence techniques [Michalski, Carbonell, and Mitchell, 1983; Baron, 1988] and *decision support systems* (DSS), to assist the designer during the decision phases of the design process. Among the most useful DSS are *multiple criteria decision making/aiding* (MCDM/A) techniques [Steuer, 1986; Vincke, 1992] and *evaluation methods* [Ettlie and Stoll, 1990].

2.2.3 Third Macroarea

The third macroarea is about detailed design activities. Tools may be divided into two classes: *computer-aided x* (CAx) and *design for x* (DFx).

Well known among CAx are *computer-aided design* (CAD), *computer-aided engineering* (CAE), *computer-aided manufacturing* (CAM), and *computer-aided testing* (CAT) [Zeid, 1991]. They typically allow a detailed design of parts and components of a new product by computer.

DFxs are a set of methodologies able to support design for product assembling, manufacturing, testing, and maintenance. Among them we remember *design for assembly* (DFA), *design for manufacturing* (DFM), *design for logistics* (DFL), and so on [Boothroyd, Dewhurst, and Knight, 1994].

2.2.4 Fourth Macroarea

The fourth macroarea picks up techniques for process design verification by means of prototypes. The first class is *rapid prototyping* [Grabowsky, et al., 1994]. It concerns a set of technologies able to directly give a physical prototype of a part starting from its drawing on a three-dimensional CAD system.

A second class is represented by *statistical experimental design* tools [Box, Hunter, and Hunter, 1978; Montgomery, 1997]. They allow the optimization of product and process parameters and performances under controlled conditions. As examples, we cite *design of experiment* (DOE) and *robust design* methods [Phadke, 1989].

In the same macroarea we find tools such as *variety reduction*, *reliability techniques*, *configuration control procedures*, *documentation management*, and *design review*.

Variety reduction aims to modularize and collect product components into families. The most important techniques in this context are *group technology* [Askin and Standridge; 1993] and *cluster analysis* [Hair et al., 1998].

Reliability techniques allow the evaluation of failing causes, effects, and critical elements of a system. They include *failure mode and effect analysis* (FMEA/FMECA) methodologies and the *fault tree analysis* (FTA) for a preventive study of potential failures [Juran, 1999; Lochner and Matar, 1990].

Configuration control procedures permit the control of issues and modifications of final design documents [ISO 9004-1, 1994, para. 8.8]. Documentation management gives the set of procedures for the management of technical documentation about the entire design life cycle [ISO 9000-1, 1994, para. 5]. Finally, design review techniques allow a formal and documented examination of the correspondence between what is specified in the design and what is really performed.

After this preliminary presentation, we may proceed to create some *relationship maps* between design activities and supporting tools. Tables 2.2 and 2.3 present two

TABLE 2.2
Map of Relationships between Project Activities and Supporting Tools for Economic–Organizational Dimension

Design Activity Legend	QFD	FCA	PM	FAM/I	DSS	RRM	VA	CGM	DR	CC	PS	DM
Market needs analysis and product features definition	●							□				
Functional analysis	□	●				□		□			□	
Explication of internal and external design activity		□	●	□		□	□			□		
Preliminary design		□										
Design parameters optimization					●		●					
Manufacturing analysis					□	□		□			□	
Design review		□		□					●			
Detailed design		●					□	□				
Engineering					□					●	□	
Design qualification												
Design changes management		□	□	□						□		●

Note: QFD, quality function deployment; FCA, functional cost analysis; PM, project management; FAM/I, financial analysis methods/investments; DSS, decision support system; RRM, risk reduction methods; VA, value analysis; CGM, creative group methods; DR, design review; CC, configuration control; PS, problem solving; DM, documentation management; ●, strong relationship; □, weak relationship.

TABLE 2.3
Map of Relationships between Design Activities and Supporting Tools for Technological Dimension

Design Activities Legend	M St.	FAST	CAx	DFx	RP	SED	DR	FMEA	FTA	VR	CC
Market needs analysis and product features definition	●										
Functional analysis		●									
Explication of internal and external design activity											
Preliminary design			●	□	□						
Design parameters optimization						●					
Manufacturing analysis			□	●		□		□	□	□	□
Design review							●	□	□	□	
Detailed design			●	□							□
Engineering				●	□	□				□	□
Design qualification											□
Design changes management						□					●

Note: M St., market studies; FAST, functional analysis and system technique; CAx, computer-aided x; DFx, design for x; RP, rapid prototyping; SED, statistical experimental design; DR, design review; FMEA, failure mode and effect analysis; FTA, fault tree analysis; VR, variety reduction; CC, configuration control; ●, strong relationship; □, weak relationship.

maps for the economic–organizational and technological dimensions, respectively. They consider two types of symbols to discriminate between strong relationships (symbol ●) and weak relationships (symbol □).

2.3 CONCLUSIONS

By comparing Tables 2.2 and 2.3 with those proposed by Pahl and Beitz [1996] we observe some remarkable differences. The complexity of a new product design and its accelerated evolution over time are the main factors responsible for this change of scenario.

Nowadays, the design methodology must give a complete answer to the problem of the contemporary definition of a product and its manufacturing process. Besides, its new horizon embraces the whole design life cycle.

Activities such as market analysis, competition evaluation, and potential customer definition are becoming integral components of design process. This new role has required an enlargement of the design team, involving differentiated professionals not easy to merge: from marketing to manufacturing, from maintenance to postsales assistance, and so on.

On the other hand, companies have progressively lost their traditional, strong vertical integration. They have opened the doors to suppliers, involving them directly in design and in its tangible outcomes.

Increasing market complexity has determined the need for a reduction of product time to market, determining a parallel path for many activities that formerly were serial.

To support this root renewal of the designer's role many efforts have been lavished toward the development of new hardware and software instruments (*technical tools*), and new operative methodologies (*organizational tools*). Particular effort was made to perform *sharable databases*, to guarantee an independent access to data, drawings, norms, and procedures.

Tables 2.2 and 2.3 show what has been done to make the design process easier and more efficient. They particularly reveal the new role played by organizational tools, formerly not included in the design process. This obviously does not mean that technological tools have stopped their growth, but that the true great novelty is about systemic or organizational supports.

Moreover, in analyzing Tables 2.2 and 2.3 we may observe that some design activities are not adequately supported, for example, the *explication of internal and external design activities* for the technological dimension, and the *design qualification* for the economic–organizational dimension.

Because of the lack of reliable empirical studies about design tools and methodologies, it is normally difficult to express a *robust* and global judgment about their effectiveness and value. It is anticipated that these studies will be performed in the near future.

In conclusion, the renewal of design activity has to be further considered and completed, and surely artificial sciences [Simon, 1981] have not yet succeeded in furnishing the promised and expected outcomes.

REFERENCES

Akao, Y. (1992), Origins and Growth of QFD, First European Conference on Quality Function Deployment, Milano, Italy.

ASI (1987), Quality Function Deployment, Executive Briefing, American Supplier Institute, Dearborn, MI.

Askin, R.G. and Standridge, C.R. (1993), *Modeling and Analysis of Manufacturing Systems*, J. Wiley & Sons, New York.

Aurand, S.S., Roberts, C.A., and Shunk, D.L. (1998), An improved methodology for evaluating the producibility of partially specified part designs, *Int. J. Comput. Integrated Manuf.*, 11(2), 153–172.

Baron, J. (1988), *Thinking and Deciding*, Cambridge University Press, New York.

Bemowski, K. (1991), The benchmarking bandwagon, *Qual. Prog.*, 24(1), 19–24.

Boothroyd, G., Dewhurst, P., and Knight, W. (1994), *Product Design for Manufacture and Assembly*, Marcel Dekker, New York.

Boschi, D., Buzzacchi, L., Calderini, M., Cantamessa, M., Paolucci, E., Ragazzi, E., and Rossetto, S. (1995), Ricerca su innovazione nella progettazione e sviluppo prodotto, Rapporto interno, Politecnico di Torino — DISPEA.

Box, Hunter, and Hunter, (1978), *Statistics for Experimenters*, John Wiley & Sons, New York.

Brassard, M. (1989), *The Memory Jogger Plus*, GOAL/QPC, Methuen, MA.

Brealey, R. and Myers, S. (1996), *Principles of Corporate Finance*, 5th ed., McGraw-Hill Series in Finance, New York.

Ertas, A. and Jones, J.C. (1996), *The Engineering Design Process*, 2nd ed., John Wiley & Sons, New York.

Ettlie, J.E. and Stoll, H.W. (1990), *Managing the Design-Manufacturing Process*, McGraw-Hill, New York.

Franceschini, F. (1998), *Quality Function Deployment: Uno Strumento Concettuale per Coniugare Qualità e Innovazione*, Ed. Il Sole 24 ORE Libri, Milano.

Franceschini, F. and Rossetto, S. (1995a), Quality and innovation: a conceptual model of their interaction, *Total Qual. Manage.*, 6(3), 221–229.

Franceschini, F. and Rossetto, S. (1995b), QFD: the problem of comparing technical/engineering design requirements, *Res. Eng. Design*, 7, 270–278.

Franceschini, F. and Rossetto, S. (1995c), Qualità, QFD e cliente: la scelta degli attributi del prodotto, *Autom. Strum.*, 43(10), 55–61.

Franceschini, F. and Rossetto, S. (1997), Design for quality: selecting product's technical features, *Qual. Eng.*, 9(4), 681–688.

Garvin, D.A. (1987), Competing on the eight dimensions of Quality, *Harv. Bus. Rev.,* 65(6), 101–109.

Grabowsky, H. et al. (1994), Support Visual Inspection with CAD — Realizing a Link at the End of the Computer Aided Process Chain for Product Development, IMS International Conference on Rapid Product Development, Stuttgart, pp. 119–130.

Hair, J.F., Anderson, R.E., Tatham, R.L., and Black, W.C. (1998), *Multivariate Data Analysis*, 5th ed., Prentice Hall, Englewood Cliffs, NJ.

Hauser, J. and Clausing, D. (1988), The house of quality, *Harv. Bus. Rev.,* 66(3), 63–73.

Hollinger, J.H. (1970), Morphologie-Idee und Grundlage einer interdisziplinaren Methoddenlehre, *Kommunikation 1*, 1, Quickborn:Schnelle.

ISO 9000-1 (1994), Quality Management and Quality Assurance Standards — Part 1: Guidelines for Selection and Use.

ISO 9004-1 (1994), Quality Management and Quality System Elements — Part 1: Guidelines.

ISO 10005 (1995), Guidelines for Quality Plans.

Juran, J.M. (1999), *Quality Control Handbook,* 5th ed., McGraw-Hill, New York.
Kahneman, D. and Lovallo, D. (1993), Timid choices and bold forecasts: a cognitive perspective on risk taking, *Manage. Sci.*, 39(1), 17–31.
Kusiak, A. and Belhe, U. (1993), Scheduling design activities, in *Information and Collaboration Models of Integration*, Nof, S.Y., Ed., Kluwer Academic, Dordrecht, NATO ASI Series.
Lochner, R.H. and Matar, J.E. (1990), *Designing for Quality*, Chapman & Hall, New York.
Mattana, G. (1994), Qualità e Misure, *De Qualitate*, 9, 7–15.
Michaels, J. and Wood, W. (1989), *Design to Cost,* J. Wiley & Sons, New York.
Michalski, R.S., Carbonell, J.G., and Mitchell, T.M. (1983), *Machine Learning*, Springer-Verlag, Heidelberg.
Miles, L.D. (1992), *Techniques of Value Analysis and Engineering*, 2nd ed., McGraw-Hill, New York.
Montgomery, D.C. (1997), *Design and Analysis of Experiments*, 4th ed., J. Wiley & Sons, New York.
Nevins, J.L. and Whitney, D.E., Eds. (1989), *Concurrent Design of Product and Processes*, McGraw-Hill, New York.
Pahl, G. and Beitz, W. (1996), *Engineering Design*, 2nd ed., Springer-Verlag, Berlin.
Phadke, M.S. (1989), *Quality Engineering Using Robust Design*, Prentice Hall International, Englewood Cliffs, NJ.
Pugh, S. (1991), *Total Design*, Addison-Wesley, New York.
Simon, H.A. (1981), *The Sciences of Artificial*, MIT Press, Cambridge, MA.
Steuer, R. (1986), *Multiple Criteria Optimization: Theory, Computation and Application*, J. Wiley & Sons, New York.
Urban, G.L. and Hauser, J.R. (1993), *Design and Marketing of New Products*, Prentice-Hall International, Englewood Cliffs, NJ.
Vincke, P. (1992), *Multiple Criteria Decision-Aid*, J. Wiley & Sons, Chichester.
Wasserman, G.S. (1993), On how to prioritize design requirements during the QFD planning process, *IIE Trans.*, 25(3), 59–65.
Wehrung, D.A. (1989), Risk taking over gains and losses: a study of oil executives, *Ann. Operation Res.*, 19, 115–139.
Zairi, M. (1992), The art of benchmarking: using customer feedback to establish a performance gap, *Total Qual. Manage.*, 3(2), 177–188.
Zeid, I. (1991), *CAD/CAM Theory and Practice*, McGraw-Hill, New York.

3 Quality Function Deployment

3.1 INTRODUCTION

The origins of quality function deployment (QFD) have not yet been exactly defined in terms of time. The general, basic concepts that are fundamental in this methodology have been known for over 40 years, even though the actual modular forms used in QFD appeared in the United States and in the Western world no earlier than 1986.

The first article to relate a short history of QFD appeared in *Quality Progress*, a magazine published by the American Society for Quality Control (ASQC) [Kogure and Akao, 1983]. The article shows that the first reports about QFD written in Japanese date back to 1967, even though before the end of the 1970s several dozen reports had been presented on the subject.

The previously mentioned article by Kogure and Akao pinpoints the official birth date as 1972, when with the help of consultants Mizuno and Furukawa engineers Nishimura and Takayanagi first developed a *quality chart* used in the shipyards of Mitsubishi Heavy Industries Ltd., in Kobe, Japan. The Kobe experiment involved the use of a matrix where the customer's requirements were listed on the page, with the columns showing the methods that had to be applied to meet these demands.

Basically the idea was that, as a result of in-depth discussions held between marketing, planning, and production, the matrix should be gradually filled in with the customer's most important requisites and with the product technical specifications expounded in the greatest possible detail. Next, various symbols were introduced to indicate whether a strong, a medium, or a weak relationship existed between the customer's requirements and the technical specifications.

Although the QFD method was extremely simple, it was hailed as a considerable step forward in respect to the hitherto virtually nonexistent aids to the design. In particular, QFD produced a galvanizing effect within the corporation in the efforts of the personnel involved to collaborate even more closely.

Two years later, Professor Yoji Akao (Deming prizewinner on QFD) founded and headed a research committee of the Japanese Society for Quality Control (JSQC) on QFD. As head of the committee he was responsible, at the end of the 1970s, for promulgating QFD as the technique used for improving the transition from design to production. Again Akao, in a successive article [Akao, 1989], declared himself to be founder of the methodology, because he was — he asserted — the first person in Japan to introduce (in 1967) the concept of QFD as a new approach to quality assurance from design right through to manufacturing. The article supplies the first operative definition of QFD as a tool in which "responsibilities for producing a quality item must be assigned to all parts of a corporation."

Even though Akao declares that he introduced the concept of QFD in 1967, Schubert ascribes to Mizuno the fatherhood of the methodology [Schubert, 1989]. According to Clausing and Pugh [1991], however, the basic ideas developed in QFD are not new, because they are rooted in *value analysis/value engineering* (VAVE), combined with marketing techniques.

The QFD diffusion throughout the United States began no earlier than 1986, almost 15 years after the experiment at the Kobe shipyards, thanks to the commitment of Don Clausing, professor at Massachusetts Institute of Technology (MIT), who was doing research work on the various ways of developing new products. At the time he was the *principal engineer* for advanced development activities at Xerox Corp., and he was first introduced to using QFD during a March 1984 visit to the Fuji Xerox Ltd. plant in Tokyo.

On his return from Japan, Clausing used his newly acquired knowledge to develop some projects at Ford Motor Co. in Dearhorn, Michigan. After that, the American Supplier Institute (ASI) organized a series of study missions in Japan aimed at focusing greater attention on the potentials and the ways of employing QFD. Now the instrument has been officially introduced to the designers' worktables in Western companies.

According to a recent definition by the ASI, QFD constitutes

> ...A system for translating customer requirements into appropriate company requirements at every stage, from research through production design and development, to manufacture, distribution, installation and marketing, sales and services [Asi, 1987].

QFD, as it has been defined, therefore constitutes a tool able to orient product design toward the real exigencies of the end user. In this sense it represents an *evident and powerful tool for laying project plans in a structured and finalized manner.* Normally, it is used before starting on the activities of development, engineering, and production of new products or services [Clausing and Pugh, 1991; Franceschini, 1993].

According to Sullivan [1996] QFD was developed as a tool contributing to the attainment of Japanese quality standards in industry. Its implementation requires the collaboration of all company staff, from top management through to workers in all the areas of a company's activities. Quality control executed in such a global manner is called company-wide quality control (CWQC).

Japanese CWQC [Akao, 1989] has contributed to enrich the American *total quality control* (TQC) approach. The new model was then accepted in the Western world with the name of *total quality management* (TQM).

QFD, therefore, represents a tool aiding TQM enabling us to avoid or at least reduce the possibility that any essential aspect of quality be neglected during the process of product design or during its revision. These concepts are effectively connected with the indications supplied by Garvin [1987], who points out that managers are often prone to neglecting one or more crucial dimensions of quality during systems design. In point of fact, quality is a *multidimensional* entity and its evaluation must necessarily involve all those characteristics that are necessary to represent it in its entirety (performance, added characteristics (*optionals*), safety, reliability, compliance with specifications, lifetime, after sales service (*service*), aesthetics, ecology, maintenance, economy of usage, etc.) [Hauser and Clausing, 1988].

3.2 INTEREST AROUSED BY QUALITY FUNCTION DEPLOYMENT

To understand the kind of results attainable with QFD, it may be interesting to name an example [Hauser and Clausing, 1988] that compares the present-day situation with that preceding the industrial revolution.

When, over 400 years ago, a knight went to a specialized blacksmith to get a new suit of armor made, the armor characteristics and design were agreed on at that time; for example, they could decide to make the armor of metal plate instead of chain mail.

The blacksmith would then transform these specifications into so many details of his production plan. He could, merely by way of example, decide on the thickness of the plate to render it less flexible; obviously this kind of decision would have had to meet the knight's approval.

Subsequently, the armorer, in considering the details of his production plan, could decide which production process would best be suited to obtain the characteristics that had been agreed on, for example, tempering the plate to harden the steel to the right point.

Finally, the armorer would determine from the production process a detailed production plan, by deciding, for example, that the fire in the forge had to be lit at 6 o'clock in the morning so that by midday it reached a sufficiently high temperature to allow him to hammer the armor into shape.

The moral of this story set in medieval times is that the definition (and *deployment*) of the armor characteristics and requisites was something extremely simple; it could be finalized by only two men: the armorer and the customer. A good deal of the process took place in the armorer's head, because he was custodian of all technical knowledge at the time.

Should we wish to reconstruct a similar situation in today's complex industrial world, we would need to be able to take customers into the plant and put them into direct contact with the workers, to have them to communicate their requirements. It does seem pretty obvious that the lifestyle found in the example is totally unfeasible in today's highly sophisticated production setting.

Nowadays, companies employ specialists having a sound technical knowledge, which has actually brought substantial advantages to end users by way of better and cheaper products, even though all this has created considerable problems in development and production processes. There again, specialists tend to shut themselves off within their specialized fields. Individually, they possess an impressive amount of technical knowledge, but there are notable difficulties in integrating them to meet customers' requests. Hence, it is necessary to develop techniques able to integrate the multiplicity of functions and so aid the two participants talking to one another, at the same time fully utilizing the enormous wealth of specific knowledge accumulated by the specialists.

The role of QFD is illustrated in the circle of company communications shown in Figure 3.1. The customer's requirements follow the circle of company commu-

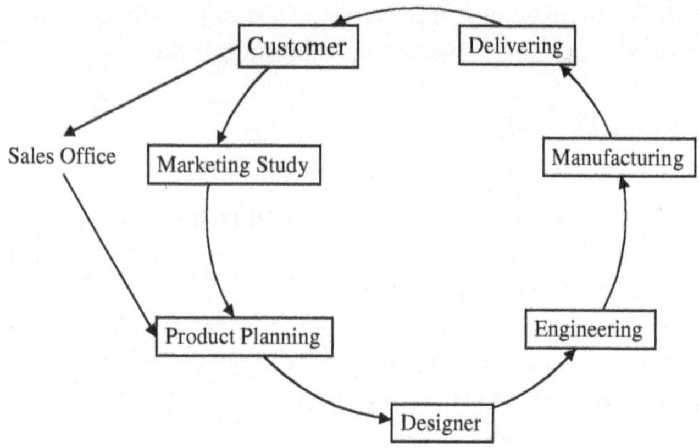

FIGURE 3.1 The circle of company communications distorts customer information.

nications and return to the customer in the form of a new product. All too often, however, in this sort of word-of-mouth communication process within a company, we find that customer requirements are not adequately translated in the passing from one function to another.

QFD is an instrument that prevents such drawbacks by having the new products pass through the various company functions, thus contributing to improvement of the company's horizontal organization.

3.3 QUALITY FUNCTION DEPLOYMENT APPROACH

The QFD process begins when we endeavor to pinpoint *customer requirements* (or *needs*), which are usually expressed in terms of qualitative characteristics, broadly defined as, for example, pleasing to look at, easy to use, working properly, safe, long lasting, stylish, comfortable, etc. During the process of product development, customer requirements are successively converted into internal company requisites, called *design specifications* (Figure 3.2).

These specifications are generally the global characteristics of a given product (usually measurable characteristics) which, if correctly developed, will have to satisfy customer requirements. Then the general specifications of the system are translated into detailed technical specifications for the subsystems or the *critical* parts (meaning those parts that will permit the realization of the essential functions constituting the reason why the product was designed).

The use of the word *parts* is considered particularly appropriate for those products that are assembled from various mechanical components. In any case, QFD can be applied just as successfully on other types of products and services in the most disparate market sectors.

Determining which *operations* are necessary for the *manufacturing* process constitutes the next step, a step often closely bound to prior capital investments in plants and machinery. Within these operational limits the manufacturing processes best suitable to attaining the desired part characteristics are established.

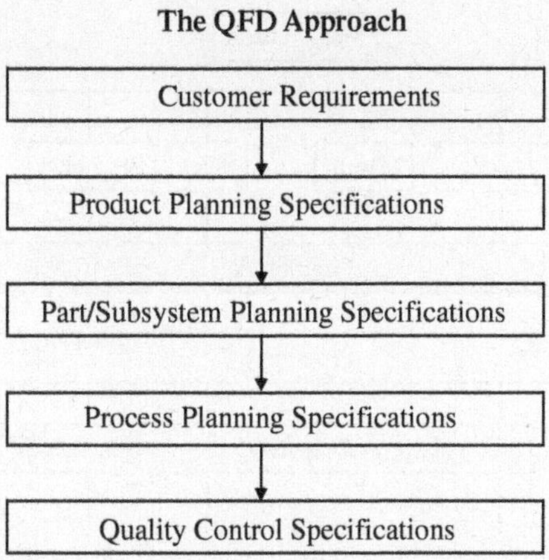

FIGURE 3.2 QFD translates customer requirements into specifications for product planning, part or subsystem planning, process planning, and quality control.

To effectively obtain the required quality characteristics, the identified manufacturing process specifications are translated into *quality control specifications*. Such specifications include, to name but a few, inspection plans for acquired materials, information needed to determine which activities will need monitoring with statistical process control (SPC), planned preventive maintenance on machinery (*total productivity maintenance* [TPM]), instructing and training operative personnel, and generally the totality of procedures and practical prescriptions in use when manufacturing a product.

This top-down (or hierarchical) approach is not, at least in appearance, dissimilar to that used by Western companies for a considerable number of years, with varying degrees of success. The differences become apparent, however, when we analyze in detail their organizational structure and their ways of dealing with customers to involve them in the product specification activities.

The structure of Western companies is usually highly pyramidal, hierarchical with rather clear backtracking reference lines. On commencement of a new project of some importance, the backtracking reference lines of many of the company functions should be widened to form the horizontal connections needed to bring the project to its conclusion. The vertical connections, however, are sometimes so strong that the corporate spirit of the various functions and the rigid respect of departmental rules form a sharp contrast to the requirements dictated by the project on hand.

The strong vertical and horizontal constraints are sometimes compared with the characteristics found in a piece of well-woven material: maximum strength of fiber, both vertical and horizontal (Figure 3.3).

3.4 STAGES OF DEVELOPMENT

From the point of view of procedure, QFD uses a series of forms called *quality tables*. The philosophy governing how QFD is to be applied is that of *management*

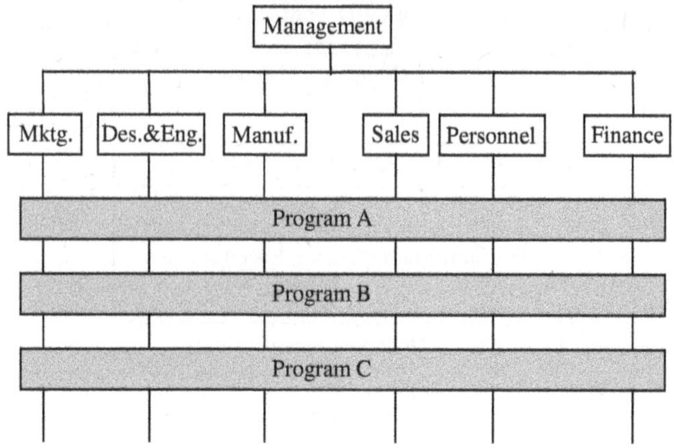

FIGURE 3.3 QFD organization. QFD helps to strengthen both the organization's vertical lines as well as the program's horizontal connections, thus improving the efficiency of the product development process. (From Sullivan, L. [1986], *Qual. Prog.*, 19(6), 39–50. With permission.)

by objectives (MBO) and *management by processes* (MBP): the emphasis is placed on both what needs to be done and how it is to be done (Conti, 1989).

Quality tables enable us to represent the variables that concur to define a given project. They show the various relationships existing among them, supplying useful indications of the levels at which they interact and of the way they interact. They consist of a series of forms with a particular layout, where the information considered important for the project development is set down. Normally, four forms are used, each one enabling the user to focus, with a varying degree of detail, on the key aspects and on the interactions occurring between the various functions.

Several different types of forms are currently in use in QFD applications [Crow, 1992; Sullivan, 1986]. They differ only in that some details may or may not be required, but the information gathered therein remains substantially equivalent.

The importance of QFD as a tool stems from the fact that both the customer and the company are compelled to make the effort to organize the project in compliance with the instructions set down in the proffered forms. As a result the documents thus obtained constitute the *common point of reference* for design revisions and successive analysis of details.

Form 1 (product planning matrix) — This compares the customer's foremost requirements (*user requirements*) with product characteristics (*product attributes*), which are the technical requisites needed to render product specifications coherent with customer expectations. The matrix thus obtained defines the relationships occurring between the two elements and their reciprocal priorities. Furthermore, it enables the user to develop comparisons between product characteristics and the best available competitor performances found on the market (*benchmarking*).

Form 2 (part deployment matrix) — This compares product characteristics with the requirements of the more important components (subsystems) into which the product can be broken down (*critical part characteristics*).

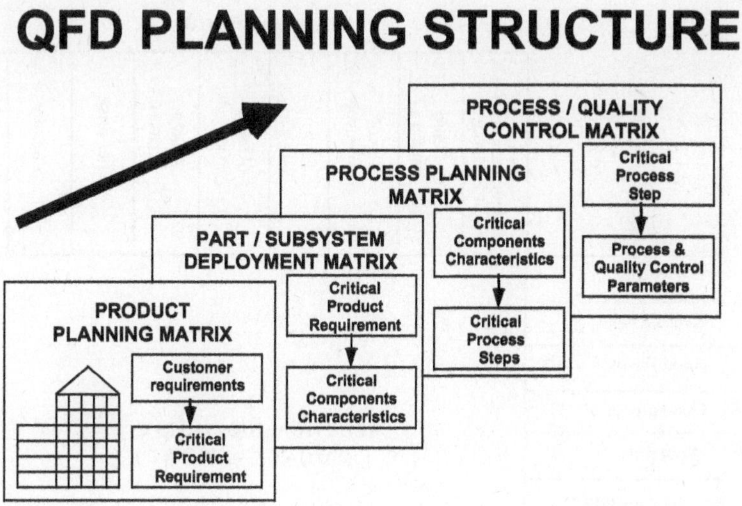

FIGURE 3.4 The logical sequence of QFD forms. The first two modules (house of quality and part characteristics) refer to product planning; the second two refer to manufacturing process planning and quality control. (From Crow, K.A. [1992], Seminar on Concurrent Engineering, DRM Associates, Rome.)

Form 3 (process planning matrix) — This relates the characteristics of the single subsystems with their respective production processes (*critical process steps*).

Form 4 (process and quality control matrix) — This defines inspection and quality control parameters and methods to be used in the production process of each process step (*quality control process steps*). In this form, in particular, each single *critical process step* is set down, as well as the relative *process control parameters*, *control points*, *control methods*, *sample size*, *frequencies*, and *check methods*. Figure 3.4 illustrates the structure as well as the logical sequence of the forms used. Besides the forms described earlier [Crow, 1992], others can be used for particular applications, for example, when the entity of the project is such that it must be necessarily broken down into a series of less complex subprojects.

3.5 HOUSE OF QUALITY

The first matrix to be used in QFD is known as the *house of quality* (HoQ). This matrix serves to describe the basic process underlying QFD: the transition (based on a strategy of input–output) from a list of customer requirements, the "what," through to a list of considerations as to "how" the requirements will be met (product characteristics).

The whats are the list of basic customer demands. These are generally rather vague requests, often expressed in imprecise terms requiring further detailed definitions. An example of a what could be the typical wish expressed by a coffee drinker: "to have a really good cup of coffee."

Customer demands, rationalized and organized according to hierarchical criteria (expectations tree) and summarized in a chart showing expected quality (*demanded*

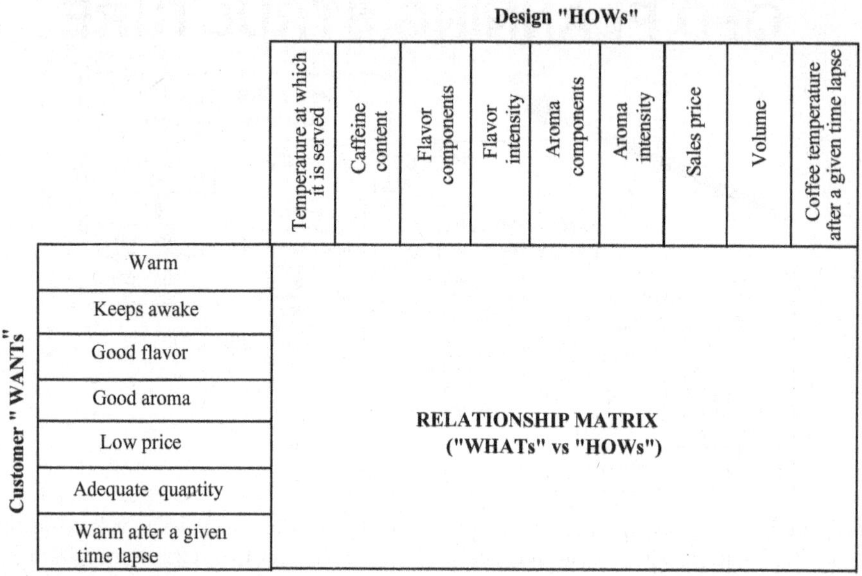

FIGURE 3.5 The whats are listed on the left of the relationship matrix. The hows are shown on the top of the relationship matrix.

quality chart) must as far as possible be kept in the customer's own words, so that they fully express the actual quality the customer asked for. The aim is to consolidate and to make available for use in successive stages of the methodology, the real as well as the latent needs as expressed by the customer, and to help in the process of transforming these needs into design specifications. The list of whats stemming from the request for a really good cup of coffee is shown in Figure 3.5.

It is necessary, at this point, to determine "how" to satisfy customer requisites, or how to meet customer expectations, from a technical point of view. Figure 3.5 also shows the technical characteristics thus identified. It is interesting to note that usually the hows impact more than the whats and that they, in turn, can reciprocally affect one another.

QFD proffers a way of unraveling this complex network of relationships through the use of a matrix, formed by hows and whats, which identifies their reciprocal relationships (*relationship matrix*).

The whats (customers' requirements or needs as defined by them) are listed horizontally on the left of the matrix, whereas the how factors (design specifications or measurable product characteristics) are shown vertically on the first line above the relationship matrix (Figure 3.5).

The relationships between the whats and the hows, that is to say the customer requirements and the measurable product characteristics, are represented by specific symbols placed at the intersections of the relationship matrix to indicate, weak, medium, or strong relationships, respectively. The symbols commonly used are a triangle for weak relationships, a circle for medium relationships, and two concentric circles for strong relationships (Figure 3.6).

Design "HOWs"

Δ: weak relationship O: medium relationship **o** : strong relationship	Temperature at which it is served	Caffeine content	Flavor components	Flavor	Aroma components	Aroma intensity	Sales price	Volume	Coffee temperature after a given time lapse
Warm	**o**								
Keeps awake	O	O							
Good flavor	Δ	Δ	**o**	O					
Good aroma					**o**	O			
Low price							**o**	O	
Adequate quantity							O	**o**	
Warm after a given time lapse	O								**o**

Customer "WHATs"

FIGURE 3.6 Specific symbols are used to indicate relationships existing between customer requirements (whats) and design specifications (hows). These unique symbols are used to define weak, medium, or strong relationships, respectively.

If no relationship is apparent, the corresponding intersections in the matrix are left blank. Rows or columns left completely blank indicate zones where the transformation of hows into whats is inapplicable. The QFD ability to transform plans into actions, due to the very fact that it induces repeated cross-checks on the various analyzed elements, makes it a particularly suitable tool for testing congruity among the various aspects involved in the definition of a project.

Parallel to the how axis, on the bottom line of the matrix, a third area is brought into focus, the axis of the "how muches." These represent the measure of the hows and are kept separate from them, because when the hows are determined, the values of the how muches are not usually known. These values will be successively determined through further analysis.

The how muches supply both a means to a guarantee that the requirements are met, and a declaration of the intended *targets* during development. Thus, they constitute specific reference values that serve as guidelines for the successive planning stage and as a means of checking progress effectively made. As far as possible, the how muches must be measurable entities, because the latter supply a greater number of opportunities to analyze and to optimize planning than nonmeasurable entities would [Kuhn, 1981].

By returning to our example, the how muches involved in our design for a cup of coffee include a definition of the following elements:

- Temperature at which it is served
- Caffeine content
- Sales price
- Amount served

- Coffee temperature after a certain given time lapse
- Factors determining the aroma, flavor, and taste requested by the customer

The process of determining the whats, hows, and how muches represents the basis for almost all QFD applications, and constitutes the lighting spark within a planning process.

3.6 ORGANIZATIONAL STRUCTURE

3.6.1 Work Team

As inferred by the work expounded thus far, QFD is meant to be developed within a work team. First of all, customer requirements and the product or service characteristics to be ultimately achieved are freely discussed; subsequently, the same information is diffused throughout the company.

The emphasis that QFD puts on teamwork results in an involvement of all company functions in the planning process, including:

- Marketing
- Design (technical management)
- Quality
- Technical assistance
- Technologies
- Production
- Suppliers

Compared with a traditional design phase review, the procedure differs: it is no longer a case of contacting only those individuals involved in the successive phase; on the contrary, everyone contributes right from the start and at every stage of product development, keeping in mind customer expectations.

To develop a project *ex novo* utilizing QFD, therefore, interdisciplinary work teams are formed, each having roughly five to seven people [Dahlgaard, Kristensen, and Kanji, 1994], embodying all the key functions mentioned earlier and having the participation, if necessary, of suppliers. The project leader of this interfunctional work team, over and above having a sound knowledge of QFD methodology, should be an expert coordinator but not constitute a domineering presence. The methodology is in fact oriented toward consensus and attains excellent results in creative work teams that run on their own, so as to allow a structured synthesis of new ideas.

3.6.2 Technical and Management Problems

The greatest difficulties that companies encounter when they try to implement QFD are organizational. QFD works best in an environment favoring innovation, and encouraging creative initiatives and sharing of information. Departmentalization along with the consequent difficulty to work in a group on projects that may last several years, on the other hand, constitutes one of the obstacles precluding implementation of QFD on a large scale.

In addition to this, often companies perceive QFD as an added workload, instead of a better way to do things. So it happens that QFD becomes submerged in a tangle of daily chores, and is ultimately perceived merely as a tool that cannot be employed because of a chronic lack of time. If companies do not integrate QFD into their daily activities, it will continue to be considered an added task.

Difficulties of a technical nature, which the method entails, are expounded in Chapter 4, giving a detailed analysis of the operative steps required in the construction of a QFD table. A few of the principal disadvantages connected to QFD usage and some risks that may be incurred in compiling the various forms follow:

- Construction of excessively long tables, which therefore become difficult to handle and to analyze
- Confusion in defining customer requisites
- Risk of mistaking product characteristics for customer requirements
- Risk of getting lost in a host of details not conformant to the operative level of intervention
- Gathering of incorrect data: often the answers given by customers are difficult to classify as needs
- Difficulty in determining the true intensity of correlation between customer needs and technical characteristics of a given product

Obviously these risks are to be kept well in mind to avoid penalizing project results.

3.7 BENEFITS OBTAINABLE FROM QUALITY FUNCTION DEPLOYMENT USAGE

According to Clausing [Eureka and Ryan, 1988], QFD was originally developed to solve three problems generally diffused in Western industry: (1) the customer's voice was held to be of no account; (2) a considerable loss of information occurred during the cycle of product development; and (3) the different interpretations were given to technical specifications by the various departments involved. Furthermore, QFD supplies the solution to two problems closely related to those mentioned earlier: the *subdivision into departments* and the *temporal serialization of activities*.

The application of QFD on a horizontal plane within the organization reduces the negative effects of departmental subdivisions. The members of a QFD team work *together* and not as separate entities.

One of the most renowned benefits of QFD is its ability to *generate and maintain involvement within the work team* over the whole product development cycle. The results of the ensuing synergy are greater than the sum of those obtained by single components. Pooling knowledge within the work team leads to improved decisional capabilities and favors the disappearance of personal prejudices [Dahlgaard, Kristensen, and Kanji, 1994].

The short-term benefits brought by QFD include shorter product development cycles, fewer modifications in planning, fewer initial problems, and improved quality and reliability.

Many companies, especially in Japan and in the United States, have benefited from QFD in that it has been instrumental in achieving notable improvements in planning cycles while at the same time attaining reduced product development times and costs. For example, Toyota Auto Body Co., Ltd., in Kariya, Japan, witnessed an overall reduction of 61% in the initial costs involved in introducing four new models of vans between January 1977 and April 1984 [Hauser and Clausing, 1988].

Furthermore, QFD contributes to the creation of a solid platform of *basic knowledge* in planning. Once the method has been successfully applied in a project, the platform of basic knowledge thus created becomes a data bank storing technical information of extreme importance. The tables and documents prepared during QFD constitute a work documentation that becomes a source of ready reference, from which to glean new and interesting ideas for future projects.

From a strictly operative point of view, QFD is best suited to attaining the following objectives:

- To define product characteristics that meet effective customer requirements (instead of presumed requirements)
- To assign, on specially structured forms, all the information deemed necessary for the development of a new product or service (a synthetic tool, albeit rich with information)
- To effect a comparative analysis of our product performances against those of competitors (comparative analysis of product profile, or *technical benchmarking*) (see Chapter 6)
- To guarantee coherence between manifest customer needs and measurable product characteristics without neglecting any point of view
- To ensure that all those in charge of each process step are constantly kept informed about the relationship between the output quality of that step and the quality of the final product
- To reduce the necessity of applying modifications and corrections during advanced stages of development, because, right from the start, everyone is conscious of all the factors that can influence project evolution
- To minimize time allotted to customer interaction
- To guarantee full coherence between product planning and planning of the relative production processes (by facilitating the integration between the various product functions and by emphasizing interactions and mutual conditionings)
- To increase the capability of a company to react, so that any errors that could stem from a faulty interpretation of priorities and objectives are kept to a minimum
- To have self-explanatory documentation on the project as it evolves
- To agree on specific reference documents, useful for the customer as well as for those involved in drawing them up, which limit to a minimum the formulation of ideas and requests that cannot be coded and, most importantly, may not find general consensus

In Chapter 4, we will analyze QFD in greater detail and see which operative steps a work team should take to organize the planning process or the replanning process that a new product or a service entails.

REFERENCES

Akao, Y. (1989), Foreword in *Better Designs in Half the Time*, King, B., Ed., Methuen, GOAL/QPC, Methuen, MA.

ASI (1987), Quality Function Deployment, Executive Briefing, American Supplier Institute, Dearborn, MI.

Clausing, D. and Pugh, S. (1991), Enhanced Quality Function Deployment, Design and Productivity International Conference, Honolulu, HI.

Conti, T. (1989), Process management and quality function deployment, *Qual. Prog.*, 22(12), 45–48.

Crow, K.A. (1992), Seminar on Concurrent Engineering, DRM Associates, Rome.

Dahlgaard, C., Kristensen, D., and Kanji, G. (1994), Break down barriers between departments, in *Advances in Total Quality Management*, Kanji, G., Ed., Carfax, Sheffield, pp. 81–89.

Eureka, W.E. and Ryan, N.E. (1988), *The Customer-Driven Company*, ASI Press, Dearborn, MI.

Franceschini, F. (1993), Impostazione di progetti di grande dimensione: il vincolo della Qualità, *Logistica Manage.*, 36, 34–42.

Garvin, D.A. (1987), Competing on the eight dimensions of quality, *Harv. Bus. Rev.*, 65(6), 101–109.

Hauser, J.R. and Clausing, D. (1988), The House of Quality, *Harv. Bus. Rev.*, 66(3), 63–73.

Hill, J.D. and Warfield, J.N. (1987), Unified program planning, *IEEE Trans. Syst., Man Cybernetics,* 2, 63–73.

Kogure, M. and Akao, Y. (1983), Quality function deployment and CWQC Japan, *Qual. Prog.*, 16, 25–29.

Kuhn, T.S. (1981), *La struttura delle rivoluzioni scientifiche*, Einaudi, Torino.

Schubert, M.A. (1989), Quality Function Deployment — A Comprehensive Tool for Planning and Development, *NAECON 89*, pp. 1498–1503.

Sullivan, L. (1986), Quality function deployment, *Qual. Prog.*, 19(6), 39–50.

4 Applying Quality Function Deployment

4.1 INTRODUCTION

The aim of this chapter is to supply a detailed description of the principal phases necessary for the construction of the house of quality (HoQ). The list of phases includes:

- Identifying customer requirements
- Identifying product and engineering design requirements
- Drawing up a relationship matrix
- Planning and deploying expected quality (by listing customer requirements in order of importance and benchmarking competitive products)
- Comparing technical characteristics (through a technical importance ranking)
- Analyzing the correlations existing between the various characteristics (correlation matrix)

Figure 4.1 illustrates the functional bonds linking operative phases and appropriate HoQ zones.

4.2 THE CUSTOMER

4.2.1 DETERMINING WHO THE CUSTOMER IS

Quality function deployment (QFD) demands that customer requirements play a leading role in planning a new product. Consequently, the first step will be to determine who the customer is, and to choose which particular type of market and end user to focus on.

In many cases, more than one customer exists, for example, the end user, the company commissioning the product, and the staff concerned with its assembly. Almost always the customer will be an insider as well as an outsider as far as the organization planning the product is concerned. Both categories must necessarily be considered; however, should a conflict arise, the customers, regarded as the outsiders, will have to be given preference because they are the ones who will buy the company's products or services [Eureka and Ryan, 1988].

Once we have determined who the customer is and have chosen which market to take as point of reference, we face the necessity of discerning its needs, be they

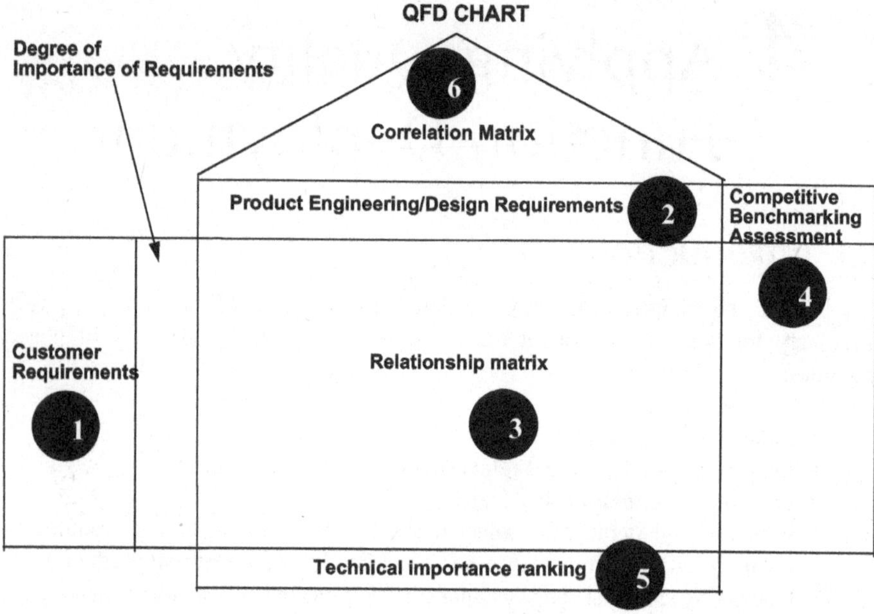

FIGURE 4.1 Main components of the house of quality (HoQ).

Data sources may be several, all to be taken into consideration: market research, specific surveys on significantly representative groups of customers–users, ad hoc questionnaires, information from marketing, technical maintenance data, complaints studies, *panels* of significant customers, *brainstorming* among company specialists, etc. [Tosalli et al., 1990]. Should the data not be sufficient, further data will have to be gathered, by contacting representative groups of customers (which may include, besides customers–users, product retailers or distributors) preferably in *vis-à-vis* or team interviews [Dahlgaard, Kristensen, and Kanji, 1994].

The *raw data* obtained from customers, also known as *source data*, constitute what has come to be defined as the *voice of the customer* (VoC), because it represents the requirements of customers–users, expressed literally in their own words (*customer verbatims*).

Many QFD specialists [Akao, 1988] prefer not to have the source data rewritten or reworded, and require only that the data be grouped according to their natural relationships, to avoid losing any of the original meaning they express.

4.2.2 Constructing the Expected Quality Table

The personnel in charge of customer input must know how to decipher those needs expressed in a somewhat vague, rather imprecise manner and write them down in their own language, utilizing carefully chosen words so that the team in charge of compiling QFD will agree on the meaning of the terms used. These terms must be kept in the customer's own words, as far as they possibly can and without creating ambiguity, because they represent the real quality requested by the customer.

TABLE 4.1
The Example Refers to a Portable Instrument for the Remote Control of a Model Aircraft

VoC	Reworded Data	Means and Remarks
"Want more than two snap roll buttons"	Easy to maneuver, Can handle difficult things	Increase snap roll buttons
"Need a neutral control on the transmitter"	Movement is stable, Can do complicated maneuvers	Add a neutral control to the transmitter

From Akao, Y. (1988), *Quality Function Deployment*, Productivity Press, Cambridge, MA. With permission.

Whenever this proves to be impossible, the team of QFD interpreters will have to rewrite the VoC into what has been defined as *reworded data* or *data expressed in word*, having single and specific implications on the product or service under study (Table 4.1). Thus, they compile a list of reworded requirements, taking care to include all those basic requirements that are often taken for granted. They make sure that the customers' likes have been identified, as well as those characteristics which, if they could be included in the product, would give the users greater satisfaction and pleasure.

Every requirement should be expressed with an adequate amount of detail, to determine their ranking order. Should the list become too long, each requirement is grouped into more generalized categories, until at most 20 or 30 requirement categories are determined.

To rationally group requests into similar categories, *affinity diagrams* or the *hierarchical cluster analysis* [Urban and Hauser, 1993], for example, may be used.

Affinity diagrams allow us to define clusters of requirements, according to the type of function they serve or to the type of problems involved, starting from the initial cluster of requisites. Clusters are formed according to team members' opinions. This procedure is often called the *KJ method*, after its inventor, Kawakita Jiro [Akao, 1988] (Figure 4.2).

The hierarchical cluster analysis, or semantic clustering, on the other hand, is based on customers' opinions. A group of customers, given the totality of requisites, is asked to create groups of similar requisites. The results thus obtained are summarized in a *co-occurrence matrix* where the generic term *ij* indicates how often, according to the customers questioned, the *i*-th and *j*-th requisites appear in the same group. By applying a *clustering* algorithm to the co-occurrence matrix, we obtain the various clusters of requisites. The reprocessed data, grouped into the same headings using one of these two methods described can be further subdivided into various subcategories or levels (typically up to three). Thus, we obtain a table of expectations, or a customer requirements tree, which is very similar to a customer satisfaction (CS) tree. This table, called a *demanded quality chart*, is placed on the first column on the left of the HoQ (see zone 1 of Figure 4.1). It shows in an organized manner the *whats* or *customer attributes* (CAs) or *customer wants/needs/requirements* set down in a rational and orderly manner according to hierarchical criteria (Table 4.2). The goal

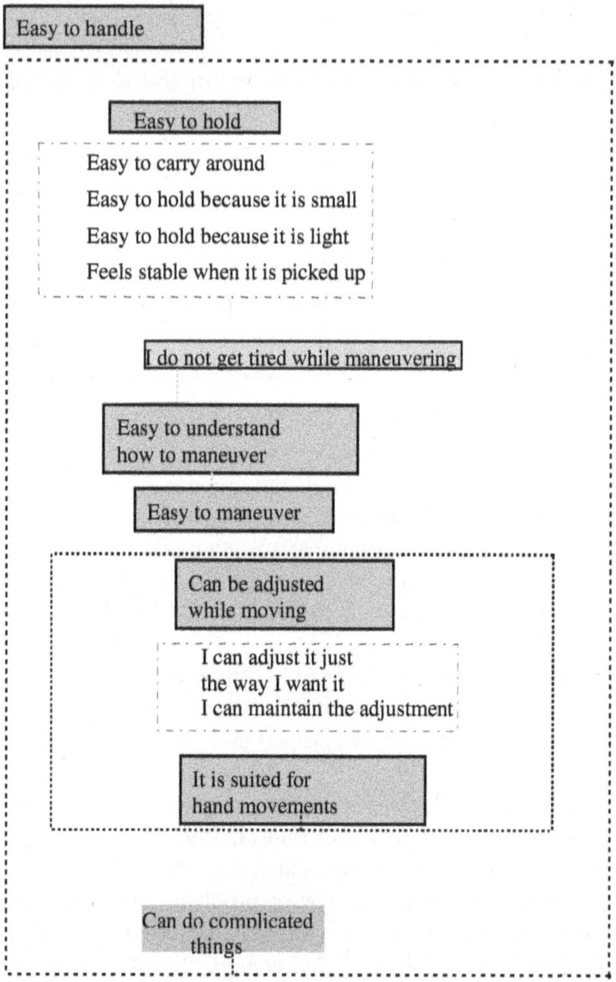

FIGURE 4.2 KJ method application to regroup word data. The example refers to a portable instrument for the remote control of a model aircraft. (From Akao, Y. [1988], *Quality Function Deployment*, Productivity Press, Cambridge, MA. With permission.)

is to consolidate and make available for successive stages of planning, all customer expressed or latent requisites.

Thus far, the first part of the construction of the HoQ has been described. It entails an extremely delicate activity concerning the outside world, our knowledge of it, and the possibility of bringing it into the company.

The measurement of market phenomena, expectations, behavior, and preferences expressed by customers, users, and consumers, is often considered as too onerous a task or even useless, because "management knows its customers well enough." In many cases, people invest in market research, study, and analysis; the results of which are, however, put to little use, hardly linked to the decisional phases. The results

TABLE 4.2
The Example Refers to the Quality Chart of a Portable Instrument for the Remote Control of a Model Aircraft

First Level	Second Level	Third Level
100 Easy to maneuver	**110** Easy to hold	**111** Easy to carry around
		112 Easy to hold because it is small
		113 Easy to hold because it is light
		114 Stable when held
		115 Stable when it is put down
	120 I do not get tired while maneuvering	**121** Has an appropriate weight
		122 Has an appropriate size
	130 Easy to understand how to maneuver	**131** Easy to understand how to use
		132 Easy to maneuver even for beginners
	140 Easy to maneuver	**141** Easy to maneuver even if it is small
		142 Easy to read the indicator

From Akao, Y. (1988), *Quality Function Deployment*, Productivity Press, Cambridge, MA. With permission.

gleaned are confined to the exclusive knowledge of single departments instead of being common knowledge throughout the company [Leoni and Raimondi, 1993].

4.2.3 Techniques Used to Determine Customer Requirements

To design a successful product it is essential that we understand potential buyers' tastes, tendencies, and commercial inclinations. An attentive analysis of customer behavior develops creativity, increases the perception of opportunities, and contributes to improve the process of decision making. What we want to determine from this type of analysis is the customer requirements; to this end several techniques have been perfected [Urban and Hauser, 1993]:

- *Personal interviews* are included in the most commonly used and most effective techniques enabling us to understand directly from customers what their needs are. Each customer, individually, is asked to describe some products that already exist on the market, how they use them, and whether any of their needs are not satisfied by these products. Interviewers insist a great deal on the needs they discover, endeavoring to define them as best they can. If, for example, customers of an automobile industry maintain that the car they bought is not comfortable, interviewers will do their best to get to the root of the meaning of the expression *not comfortable*, illustrating it with the greatest number of details. Interviews produce the raw material on which to work. Work teams listen to the interviews, transcribe them, and endeavor to determine all the needs expressed by the customer — even those implied or not quite evident. Generally, it has been found empirically that 20 or 30 interviews [Urban and Hauser, 1993] are sufficient to glean most customer requirements. Figure 4.3 shows the relationship between the number of customers interviewed and the percentage of requisites determined for any long-lasting article. It is important, when analyzing the interviews, that individual members of the work group have their own perception of the requisites, whether or not explicitly expressed. This procedure makes it virtually impossible for any factors to go unheeded, because the number of persons involved in the work group analyzing the interviews always exceeds one.

- *Focus groups* include six to eight customers requested to talk about their requirements. This has various advantages: the fact that one person in the group makes a statement provokes different reactions and comments among the other persons in the group, and this allows us to clarify any intuitive sensations. During a conversation lasting about 2 h, a customer talks on an average for more or less than 15 to 20 min, compared with an hour in a personal interview. During these conversations, the manager monitors the group using a two-way mirror or a camera, studying group reactions and behavior.
- *Structured qualitative techniques* are used where customers are requested to make some considerations concerning the products, examining them in groups of threes. They are asked to choose which two products are most similar and to say why; the same is done with the two most different products, so as to establish a relationship between the products.
- *Product analysis techniques* involve asking customers to say out loud how they buy, use, describe, and evaluate a given product. Their statements are recorded on tape and subdivided according to criteria of method, cause, and aim, allowing us to identify all the factors that may in some way contribute to the planning process.

In other cases, where a company may lack the resources or the time required for developing qualitative techniques, other expedients are used. For example, the product is shown in public places and customers are allowed to freely examine it and try it. The technical staff members who designed the product remain in the immediate vicinity of the product being exhibited, and record all the comments made by the public.

In some companies, the work team intent on planning a new product will often draw up an initial list of customer attributes on the basis of members' experience.

4.2.4 Product Perceptual Maps

A company wishing to create a successful product must know how similar products already existent on the market are perceived by customers. To have some kind of perception about a product means to be able to evaluate it in terms of quality, cost, and utility with respect to other existing products. The placing on the market of a new product is a delicate task because, over and above customer satisfaction, we must keep in mind the products of competitors. [Urban and Hauser, 1994; Eureka and Ryan, 1988].

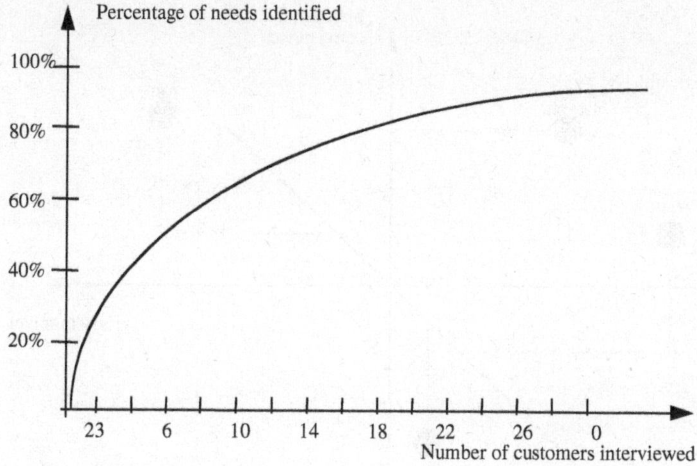

FIGURE 4.3 Percentage of needs identified vs. number of customers interviewed for a generic lasting article. (From Urban, G.L. and Hauser, J.R. [1993], *Design and Marketing of New Products*, Prentice Hall, Englewood Cliffs, NJ. With permission.)

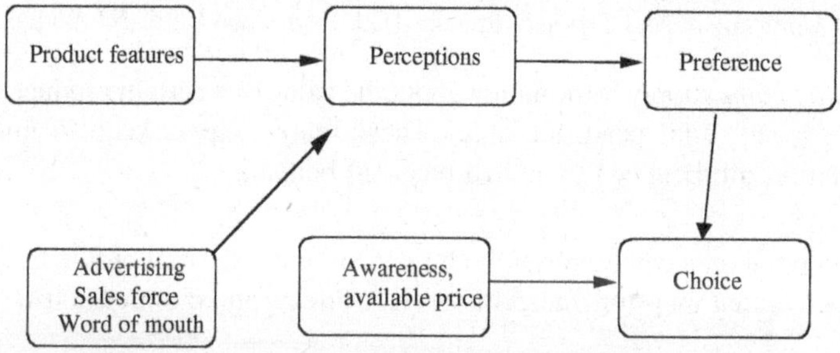

FIGURE 4.4 Brunswik lens model. (From Brunswik, E. [1952], *The Conceptual Framework of Psychology*, University of Chicago Press, Chicago. With permission.)

Brunswik's model [Brunswik, 1952; Urban and Hauser, 1993] maintains that consumers create their own opinions, which makes them prefer one certain product to another, exclusively on the basis of subjective perceptions. Customers use their perceptions as lenses to filter the complex web of messages that are transmitted to them through various communication and persuasion channels (Figure 4.4).

Some typical instruments used to determine how a product ranks on the market, in respect to the benefits it produces, are:

- *Perceptual maps* summarize the dimensions according to how a customer perceives, judges, and identifies one product compared with another. To be able make this kind of map the company must know the number of dimensions to be considered for evaluation, the type and the needs underlying the dimensions, where the competitors stand, and what margins of improvement can be reached.

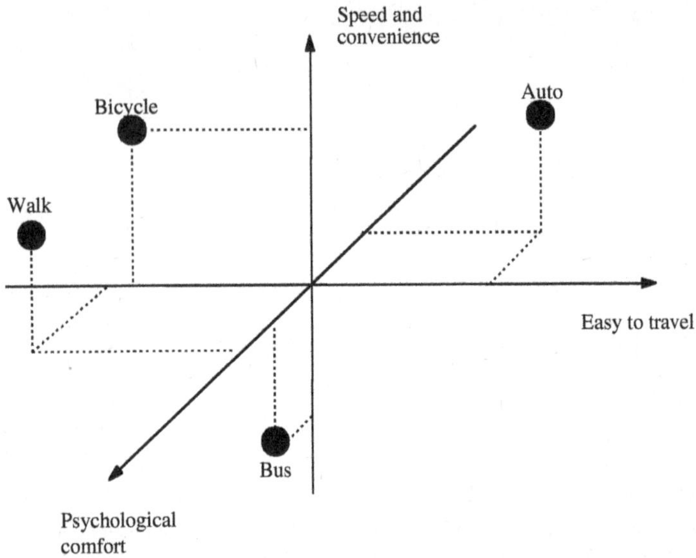

FIGURE 4.5 Perceptual map of transportation services. (From Urban, G.L. and Hauser, J.R. [1993], *Design and Marketing of New Products,* Prentice Hall, Englewood Cliffs, NJ. With permission.)

- *Value maps* supply information about the value of a certain product, a kind of "benefit unit per price unit." These maps help to keep in mind the relationship between price and received benefits.

A perceptual map can represent both services and industrial products. Figure 4.5 shows a perceptual map for transport services having three dimensions:

- *Speed and convenience* reflect the method of a punctual transport service that takes its passengers to their destination quickly, which is available whenever needed, allowing complete freedom of movement.
- *Ease to travel* implies correct temperature, absence of problems connected with bad weather, ease with which luggage is transported, etc.
- *Psychological comfort* includes the possibility of relaxation and the absence of preoccupations about being robbed, hit, or bothered.

In this example [Urban and Hauser, 1993] only three dimensions are represented; the number may, however, vary according to the type of survey on hand. The preceding three dimensions are considered classic examples of primary customer needs; each one can, in turn, be expanded into a series of secondary needs, each of which can be expanded into tertiary needs, and so on.

Within a process of innovation, the result that can be achieved using these maps is dual: it causes product repositioning with respect to the same dimensions, and identification of new dimensions or market segments.

When no opportunities for improvement are even dimly seen on the market scene, the company could decide to add a new dimension, if it has identified the requirements that still need satisfying. All too often, the action a manager may regard as revolutionary when conceiving a new product is hardly perceived by the customer; for this reason a constant check must be kept on the customer's reactions, to be able to understand what is really needed.

4.2.5 Evaluating the Importance of Attributes

Perceptual maps show the ranking of a product on the market, but are not able to suggest which dimensions ought to be awarded greater attention.

To understand which factors are more important for the customer, we could place on the market a series of different products, and evaluate buyers' reactions. This method, however, is costly, often impractical, and decidedly time consuming.

Thus, various methods were developed to measure the importance that customers assign to each single attribute of a product. The first method consists of directly evaluating the importance of a list of attributes, asking customers to express the weight they think they ought to assign to each element, by filling in a relevant questionnaire (as we shall see later). This activity usually involves the use of some particular qualitative evaluation scales. The terms used on the scales can be changed according to how the questions are stated and according to the type of information we expect from the customer. Normally, it is difficult to establish which evaluation scale is best suited to show a buyer's opinion; this is mostly based on perceptions and sensations hardly interpretable or quantifiable in numerical values.

Another method consists of asking individual customers to express their evaluations using a numerical value, for example, from 1 to 5, for each attribute identified. Here we endeavor to assign a reference value or to explain beforehand what type of opinion is associated with each value.

Sometimes the procedure is more methodical: customers are asked to assign a value of 10 to the attribute they consider most important; and then they are asked to assign values to the other factors, possibly working down the scale from the most to the least important factor.

Another sort of evaluation is that of allocating 100 points across the five dimensions according to how important each determinant is perceived. The customer is given a total number of points to divide among all the dimensions. The advantage of this method is that the customer must keep in mind the possible trade-offs between the various attributes.

In conclusion, the number of times a customer mentions a certain attribute in a survey is to be noted. Empirical checks on questionnaires have proved that the reasons for customer satisfaction and dissatisfaction after buying a certain product appear at the top or at the bottom of the list of attributes, respectively, compiled by the customer.

An important aspect to consider when gathering and processing data is that individual customers have their own reference system and that, consequently, the aggregation of various judgments may not occur immediately due to the different meaning each individual attributes to the levels of the scales used.

TABLE 4.3
Extracting Technical Characteristics

Required Item (Third-Level Items)		Technical Characteristics
111 Easy to carry around	→	Weight, dimensions, shape, portability
112 Small enough to carry around easily	→	Dimensions, shape, portability
113 Light enough to carry around easily	→	Weight, shape, portability
114 Feels stable when held	→	Weight, gravity center, angle of incline
115 Stable when set down	→	Shape, gravity center, stability
116 Even beginners can operate easily	→	Location of buttons, sensitivity to touch
....	→	....
141 Can be operated easily, even though small in size	→	Weight, shape, effort needed to move stick, stick sensitivity to touch, strength needed to hold lever in position, location of buttons, location of knobs, effort needed to operate knobs, knobs sensitivity

From Akao, Y. (1988), *Quality Function Deployment*, Productivity Press, Cambridge, MA. With permission.

4.3 DETERMINING TECHNICAL CHARACTERISTICS

As our first step, strictly within the domain of marketing, we determine what the customer requires and then what should be done. As our second step, more markedly within the domain of technical planning departments, we decide how to obtain the desired result. To achieve this goal the interfunctional QFD team members must, starting from customer needs, determine the measurable and controllable product characteristics involved in design that will enable them to reach an exhaustive evaluation of the product or service.

This is a particularly exacting step because it implies translating the market model as expressed in subjective terms by the customer's words, into objective factors of a technical nature (*performance characteristics*), that is, into a description of the product or service expressed in the designer's own language (the so-called *voice of the engineer* [VoE]). Thus, a list is compiled showing the *technical design requirements, characteristics, and parameters* or the *engineering characteristics* (ECs) that represent the *hows* determined by the engineer. Some authors call these parameters *substitute quality characteristics* (SQCs) because they substitute customer requirements and constitute the input data for design (Table 4.3).

At least one EC should be identifiable for each customer request, even though each single EC may affect more than one customer request. If an EC does not affect any customer request it may be redundant, or the QFD team may have forgotten to include a specific customer requirement. On the other hand, a customer requirement not influenced by any technical attributes found on the list constitutes a new opportunity to further study the technical and functional characteristics of the product.

The EC should proffer a description of the product or service in measurable terms (offered quality) and should directly affect customer perception concerning quality. For example, in a car, the mass of the door — expressed in kilograms — is one characteristic that the customer will effectively feel, therefore it will be a noteworthy EC. On the contrary, the thickness of the sheet metal used on the door is a very important technical characteristic, even though the customer most probably does not directly perceive it. Its thickness actually influences the customer only in that it affects the weight of the door and other technical characteristics such as resistance to deformation in the event of collision [Hauser and Clausing, 1988].

As with the customer requests, to obtain as precise a description as possible, the design characteristics are also grouped into a first, second, and third level (*it is important to ensure that at least the third-level ECs are all quantifiable*).

In QFD applications in the service sector, usually the term *technical characteristics* is substituted by *quality elements* to indicate such characteristics as kindness or courtesy, so it is difficult to identify a unit of measure. The same term is used also in the higher level ECs, for example, shape, which is an element of quality that might include the specifications at lower levels of height or depth as measurable quality characteristics.

At the end of this second step we arrive at the identification of the quality characteristics, or the EC tree, which is represented by the columns in the HoQ (see area 2 of Figure 4.1).

4.4 CREATING THE RELATIONSHIP MATRIX

The interfunctional team's successive task is to fill in the body of the HoQ, constituted by the so-called *relationship matrix* (see zone 3 of Figure 4.1), which indicates how the technical decisions affect the satisfaction of each customer requirement.

For each element in the matrix, we try to obtain an answer to the question: To what extent can the technical characteristics of a product or service (determined in step 2) affect the quality expected by the customers in terms of their degree of satisfaction?

The team discusses the answers to these questions until a consensus is reached. The agreement on the evaluations is based on former experiences in the technical field, on the customer's responses, and on the data obtained through statistical analysis. The relationships between the requirements and the characteristics are expressed in a qualitative manner, or at most in a semiquantitative manner, by the factors of *correlation intensity* r_{ij}, for example, strong, medium, weak, doubtful, or nonexistent; and are coded using letters, numbers, or specific conventional symbols placed at the intersections of the matrix.

The symbols commonly used are a triangle for weak relationships, a circle for medium relationships, and two concentric circles for strong relationships (Figure 4.6).

◎ : strong relationship	✔ : positive strong relationship
○ : medium relationship	✓ : positive medium relationship
Δ : weak relationship	× : negative medium relationship
	✖ : negative strong relationship

FIGURE 4.6 Symbols used to construct a production planning matrix. (From Akao, Y. [1988], *Quality Function Deployment*, Productivity Press, Cambridge, MA. With permission.)

A strong correlation implies that a small variation (either positive or negative) in the value of the j-th indicator of technical efficiency (the EC, ec_j) may produce a considerable variation (whether positive or negative) in the degree of satisfaction $gds(ca_i)$ of the i-th need (which is the *customer requirement* ca_i).

If we consider that the degree of satisfaction of an i-th customer need depends on the values assumed by the set of m ec_j, which describe the product in technical terms, we may write:

$$gds(ca_i) = f(ec_1, ec_2, \ldots, ec_j, \ldots, ec_m)$$

where f is the implicit function of m variables. Thus, we can define analytically the factors of correlation intensity (assuming that f is derivable) as:

$$r_{ij} = \left| \frac{\partial[gds(ca_i)]}{\partial(ec_j)} \right| \geq 0$$

Makabe's version of QFD (also presented by Hauser and Clausing, [1988]) distinguishes the positive relationships (strong or weak positive relationships) from the negative relationships (strong or weak negative relationships) (see Figure 4.6). In this case the correlation intensity r_{ij} is, for example, a strong negative correlation if an increase — however small — in the value of ec_j produces a considerable fall in $gds(ca_i)$, and vice versa a decrease in ec_j produces a considerable increase in $gds(ca_i)$. In this case, the factors of correlation intensity may be defined as:

$$r'_{ij} = \frac{\partial[gds(ca_i)]}{\partial(ec_j)} \gtrless 0$$

If no correlations exist, the corresponding intersections in the matrix are to be left blank (which means $r_{ij} = 0$).

4.5 EXPECTED QUALITY DEPLOYMENT

After having analyzed individual customers' requests, some of these are selected to ensure that the new product will immediately give them a greater degree of satisfaction.

This process, called *quality planning* [Akao, 1988], is based on the classification and prioritization of customers' expectations and on the comparative analysis from the customers' points of view.

4.5.1 Customer Needs and Kano's Model

Not all customer expectations are equally important: we are able to verify how the introduction (or, vice versa the lack of) of a quality attribute requested by the customer for a certain product or service will result in quite a different degree of satisfaction (or dissatisfaction), depending on the type of attribute.

As evidenced by Kano et al. [1984], the relationship between the physical quality level of a product or service and the degree of customer satisfaction — which once upon a time was considered to be a linear, or one-dimensional, type of relationship — is in effect multidimensional.

According to research conducted by Kano, actually, the quality attributes defined by the customer may be subdivided into five categories:

- Type B *(basic)* attributes, or *must be* or *expected*
- Type O (*one-dimensional*) attributes
- Type E (*excitement*) attributes
- Type I (*indifferent*) attributes
- Type R (*reverse*) attributes

The first attributes are those that derive from basic needs; often remaining implicit for the user, they are part of the so-called *expected quality.* The presence of these attributes does not constitute, in the eyes of the customer, a source of satisfaction or of dissatisfaction, but their absence results in strong dissatisfaction. Therefore, an increased level of fulfillment for this type of requirement does not lead to a higher degree of overall satisfaction; vice versa, a decrease in the degree of fulfillment of these requirements leads to a veritable "plunge" in the customer's overall satisfaction.

The O-type attributes (linear, or one-dimensional, attributes) are product characteristics that the customer does not consider to be particularly captivating. Their presence contributes to increasing the customer's satisfaction and their absence produces a certain amount of dissatisfaction. If we increase or decrease to the same extent the degree of fulfillment of these needs, we obtain an increase or a decrease, respectively, in the customer's global levels of satisfaction, the variations being more or less proportional.

The E-type attributes attract and delight the customer and are those that contribute to differentiating a particular product from that of competitors. Their presence — or an increase in their level of fulfillment — produces a high level of contentment on the part of the customer, whereas their absence does not produce any particular level of dissatisfaction.

The I-type attributes are those whose presence or absence produces neither satisfaction nor dissatisfaction in the customer. The R-type attributes, on the other hand, are those whose presence produces dissatisfaction in the customer, whereas their absence is a source of satisfaction.

Naturally, in an analysis of customer requirements, we may face attributes belonging to all five categories just listed. It must be remembered, however, that *indifferent*-type attributes are not to be taken into account, because they constitute that extra effort by the company that produces no satisfaction in the customer.

When dealing with reverse-type attributes, these must be transformed into their dual one-dimensional counterparts because QFD operates within the field of *positive quality,* meaning that it deals with all the characteristics of a product or service able

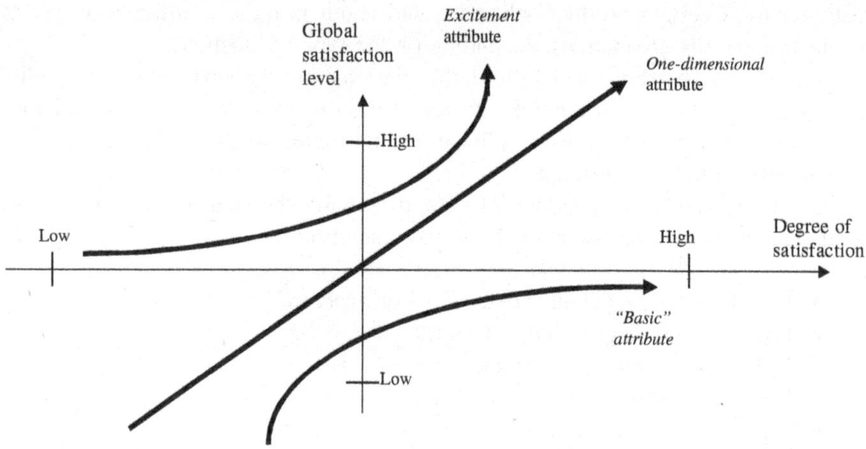

FIGURE 4.7 Kano's model. (From Kano, N. et al. [1984], *J. Jpn. Soc. Qual. Control,* 14(2), 39–48. With permission.)

to increase value for the customer in terms of adequate usage [Conti, 1992]. Consequently, the expected quality table will only show the attributes belonging to categories B, O, and E (Figure 4.7).

The three categories thus determined are not static, they are dynamic in time, in the sense that the excitement attributes tend to become one-dimensional, whereas the one-dimensional attributes tend to become basic.

We must be quite clear about this distinction propounded by Kano when we prepare a quality table. The distinction will differ depending on the company, but its presence will reflect on the way the customer requests are prioritized.

In a scenario of ever-increasing competitiveness between companies producing new products, it is easily foreseeable that the number of products designed to satisfy E-type requests will increase. This implies an ever more frequent recourse to technological innovations (see Chapter 1). The determination and analysis of E-type requirements — right from the very first stages of the project — are those that allow the company to foresee the presence of possible technological bottlenecks as well as the feasibility of propounded solutions.

4.5.2 Prioritization of Customer Requirements

Customer requirements, which belong to different categories according to how much they affect the degree of global satisfaction of a product or service, must be ranked according to the customer's system of preferences. This is an extremely delicate step requiring the correct implementation.

Let us consider two types of customer requirements concerning the doors of a sports car: they must be easy to lock and the electric windows must be fast and easy to use. Immediately we encounter the problem that to increase the speed of closure, a more powerful electric motor will have to be installed and it will occupy more space. This, however, would make the door heavier and more difficult to close [Hauser and Clausing, 1988]. Sometimes a solution to equally satisfy both requirements can be found, but usually designers are forced to reach a compromise, or trade-off, between the various requests made by the customer. To be absolutely certain that the customer's voice, and not the designer's, bears more weight in these matters, it is necessary to determine the relative importance (or customer priority) of each requirement as determined by the customer.

The prioritization of product characteristics is normally carried out on the basis of the team members' direct experience with the customer (the procedure is not recommendable, even though in some cases it produces very good results) or on the basis of surveys. In some sectors statistical techniques are employed that allow customers to determine their preferences with respect to products already existing or even with hypothetical products.

Other companies, instead of direct confrontations, use techniques of *revealed preference (preference regression and conjoint analysis)* [Urban and Hauser, 1993], which judge customer tastes on the basis not only of their expressed preferences (e.g., in questionnaires) but also of their actions. For example, consumers may indicate, in questionnaires, that it is important for breakfast cereals to contain no sugar, but their market actions may not exactly reflect their requests. This is in effect a more costly and a more difficult approach to implement, but it does give more accurate responses.

The traditional QFD methodology solves this delicate problem of assigning degrees of priority to customer requirements by ranking them according to a scale from 1 (for a requisite of negligible importance) to 5 (for an indispensable requisite) or from 1 to 10 (see Section 4.2.5). This prioritization, or *relative-importance ranking*, is obtained by drawing up questionnaires (Figure 4.8) that require customers to prioritize a list of predetermined requisites (using the same type of conventional scale from 1 to 5 or from 1 to 10).

The results of these questionnaires are analyzed to find the statistical distribution of the weights over each customer requirement. If the distribution is manifestly unimodal, the mean values of the evaluations are taken into account. If, on the other hand, the distribution is, for example, bimodal, with significant variations in importance levels, it does not make much sense to calculate the average scores, because this sort of distribution reveals that we are probably confronting two different market segments.*

Some authors [Sanchez et al., 1993], once customer requirements have been classified into B, O, or E categories, according to Kano's classification, take the next step that consists of assigning scores according to their importance. For example, a score of 3 will be assigned to B-type requests, whereas O-type requests will score 4 and E-type requests will score 5. By using this method, the excitement-type attributes are obviously given greater resonance, but the customers' opinions are in fact distorted. Although it may be true that the E-type expectations are more markedly relevant in global customer satisfaction, it is by no means true that B-type requests are less important, because if these are neglected the customer will be merciless.

* According to Kotler (in Hauser and Clausing, 1988), the segmentation of the market is "the process by which the market is divided into many distinct groups of consumers who may require a different marketing-mix of products."

Two questions will be asked of you. The answer in column 1 indicates how important each item is in influencing your purchase decision. The answer in column 2 asks you to evaluate each manufacturer on each item, after you have tried each one.

Question 1: The items listed here may influence your purchasing decisions for a radio-controlled product. In column 1, please rank how much influence these items have on your purchase decision. Please circle the appropriate level.

Question 2: Whose radio control do you currently own? Please fill in the name of the manufacturer.

Company X..........name of manufacturer ( )
Company Y..........name of manufacturer ( )
Company Z..........name of manufacturer ( )

In column 2, please evaluate each manufacturer's product after using it. Please circle the appropriate level.

Items to judge the product	No influence at all / Minor influence / Some influence / Strong influence / Very strong influence	Very bad / Bad / Fair / Good / Very good
(example) easy to hold	1 2 3 4 5	X 1 2 3 4 5 Y 1 2 3 4 5 Z 1 2 3 4 5

FIGURE 4.8 Example of questionnaire used to construct customer requirements prioritization. (Akao, Y. [1988], *Quality Function Deployment*, Productivity Press, Cambridge, MA. With permission.)

According to other authors [Akao, 1988; Armacoast et al., 1994], a more adequate approach to the prioritization of customer requirements would involve the use of the *analytical hierarchy process* (AHP) (see Chapter 5).

4.5.3 Benchmarking on the Basis of Perceived Quality

To deal correctly with each of the requirements expressed by customers, it is useful, whenever possible, to have customers compare the product produced by their own company with competitive products belonging to the same market segment. To be able to do this, by way of example, the following method may be adopted. The same questionnaire is sent to a group of customers to inquire on the level of importance of each requirement (see Figure 4.8). In it individual customers are requested to evaluate the degree of satisfaction obtained within their own company from the use of the product, as well as the degree of satisfaction obtained from the product marketed by their strongest competitors. The scale, from 1 (very bad) to 5 (very good), is the same as the one used to evaluate the importance of the requisites affecting a customer's decision to buy. This phase is called the *analysis of competitiveness from a customer point of view* or *benchmarking on the basis of perceived quality.*

4.5.4 Target Values of Expectations

On the basis of the values obtained with the procedures previously described and on the basis of an analysis of the priority levels assigned to the various requirements, the sales characteristics or *strengths* of a given product can be established.

By way of example, if we were to construct a quality chart for planning a pencil (Figure 4.9), requisite 3 — "point lasts" — has been given a 5 score (indispensable) as a requisite affecting a customer's decision to buy it — column A, "level of importance."

The evaluation of the satisfaction of this requirement by the company's present product — column B, present model — is 4 (satisfied), whereas the product marketed by the other two major competitors — in the respective columns "competitor X" and "competitor Y" — have obtained a 5 score (requisite fully satisfied) and a 3 score (requisite satisfied to some degree), respectively. This indicates that the satisfaction of this requirement constitutes one of the product's potential strengths of obvious importance for an improved brand image. A conventional score of 1.5 is assigned to the very important strengths in the relevant column (column E), whereas for those requisites whose satisfaction is considered as a possible strength, the assigned score is 1.2. Those requirements not considered as strengths bear a weight of 1.

In developing a quality plan for a given product, it is also necessary to fix *target values* for satisfying requirements, keeping in mind both the company strategies and the values obtained from analyzing the competitive products. These target values go in the column named "targets for new model" in the HoQ (column C in Figure 4.9), utilizing the same 1 to 5 scale used in the benchmarking analysis.

The values in column D — ratio or degree of improvement or upgrading factor — represent the measure of improvement necessary to reach target values. These are worked out by calculating the ratio between the target value and the customer's evaluation — column B, "present model."

For example, for requisite 3 — "point lasts" — the target value assigned to it is equal to 5 (the highest), in consideration of the fact that this requisite represents an obvious strength, and in this case our company has a lower performance level than its competitors do.

Thus, the improvement ratio for satisfying that requisite will be:

$$\text{Targets new model/present model} = 5/4 = 1.25$$

From an operative point of view these can be utilized as aids to strategic decision making and to company policy concerning the new product. Both these elements (customer's voice and company policy) are taken into account, by calculating the so-called "absolute weight" of the requirement (column F). The calculation is as follows:

	A	A'	B			C	D	E	F	G
			Benchmarking on the basis of Perceived Quality			Quality Planning				
Customer Requirements	Degree of importance	Relative importance	Present model	Company X	Company Y	Targets for new model	Improvement ratio	Strength	ABSOLUTE WEIGHT	RELATIVE WEIGHT
1) Easy to hold	2	17%	4	4	4	4	1	1.0	2.0	11%
2) Does not smear	3	25%	5	4	5	5	1.00	1.2	3.6	20%
3) Point lasts	5	41%	4	5	3	5	1.25	1.5	9.4	53%
4) Does not roll	2	17%	3	4	4	4	1.33	1.0	2.7	15%
Total	12	100%						Total	17.7	100%

LEGEND

D = C / B F = A * D * E

Requirement Importance

1. Not important at all
2. Minor importance
3. Some importance
4. Strong importance
5. Very strong importance

Benchmarking on the basis of perceived quality

1. Very unsatisfied
2. Unsatisfied
3. Relatively satisfied
4. Satisfied
5. Very satisfied

FIGURE 4.9 Deployment of the quality chart for a pencil. (From Wasserman, G.S. [1993], *IIE Trans.*, 25(3), 59–65. With permission.)

$$\text{Absolute weight} = \text{Level of importance} \cdot \text{Improvement ratio} \cdot \text{Strength}$$

For example, the absolute weight P_{r3} of the requisite "easy to hold is":

$$P_{r3} = 5 \cdot 1.25 \cdot 1.5 \cong 9.4$$

The indications resulting from the quality planning model constitute the input to the communicative–persuasive channel (see Chapter 1) employed in imaging strategies for the new product (modification of customers' expected and perceived quality).

4.6 TECHNICAL COMPARISON

4.6.1 EVALUATING THE IMPORTANCE OF CHARACTERISTICS

The relationship matrix relates customer requirements with product characteristics by using, as coefficients, the symbols of the ordinal scales in Figure 4.6.

By using the information contained in the relationship matrix we can, taking as the starting point the priorities assigned to the customer requirements, determine a list of levels of importance assigned to product characteristics. The importance of each characteristic is evaluated on the basis of:

- The importance of the customer requirements to which it is correlated
- The level of correlation
- The degree of difficulty its realization entails

The classic method (*independent scoring method*) used to rank the technical characteristics of a product requires two operative steps [Akao, 1988]. The first step consists of converting the relationships expressed in symbols between customer requirements and the product characteristics into equivalent numerical values. This conversion from an ordinal to a cardinal scale utilizes 1, 3, 9, or 1, 3, 5, or 1, 5, 9 scales (using the symbols in the left-hand column of Figure 4.6). The relationships are coded using a -2,-1, 1, 2 scale if we use the symbols in the right-hand column of Figure 4.6. To convert ordinal evaluations on three levels, the standard system of weights 1, 3, 9 is more commonly used, even though alternative systems can be utilized (1, 2, 4, or 1, 3, 5). This is how the numerical coefficients $r_{i,j}$ in the relationship matrix R are generated.

The second step entails determining the level of importance w_j of each technical characteristic. It is obtained by summing the products of relative importance of each customer requirement (column A′ in Figure 4.9) multiplied by the *quantified* value of the relationship existing between that j-th characteristic and each of the requirements related to it. We obtain:

$$w_j = \sum_{i=1}^{n} d_i \cdot r_{i,j} \tag{4.1}$$

where d_i is the degree of relative importance of the i-th customer requisite, $i = 1, 2, \ldots, n$; $r_{i,j}$ is the cardinal relationship between the i-th customer requisite and the j-th product characteristic, $i = 1, 2, \ldots, n; j = 1, 2, \ldots, m$; w_j is the technical importance rating of the j-th characteristic, $j = 1, 2, \ldots, m$; n is the number of customer requirements; and m is the number of product characteristics.

In vectorial form, we can write:

$$w_j = \mathbf{d} \cdot \mathbf{R}_j$$

where **d** is the row-vector of the degree of relative importance attributed to the customer requirements and $\mathbf{R}_j$ is the *j*-th column of the relationship matrix.

The measure of the (absolute) level of technical importance may be transformed into a measure of relative technical importance w_j^*, expressed as a percentage:

$$w_j^* = \frac{w_j}{\sum_{j=1}^{m} w_j}$$

$$j = 1, 2, \ldots, m$$

The latter represents the importance that the customer indirectly assigns to each product characteristic and may be used to define a ranking order of the levels of attentiveness the designer will have to attribute to the technical engineering characteristics during design.

Should we, during the prioritization process, wish to take into account not only the degree of relative importance d_i assigned by the customer to each of the requisites but also the relative weight D_i calculated on the basis of company policy (column G in Figure 4.9), we are able to calculate the *absolute weight* W_j of the *j*-th characteristic in the following manner:

$$W_j = \sum_{i=1}^{n} D_i \cdot r_{i,j} \tag{4.2}$$

Therefore, the *relative normalized* weight W_j^* of the *j*-th technical engineering characteristic will be:

$$W_j^* = \frac{W_j}{\sum_{j=1}^{m} W_j}$$

$$j = 1, 2, \ldots, m$$

Let us now fully reconsider the example shown in Figure 4.9, concerning the planning of a pencil [Wasserman, 1993] to make clear the described procedure (Figure 4.10). From a market research the following customer requirements were obtained:

- Easy to hold
- Does not smear
- Point lasts
- Does not roll

The planning team, on the basis of these requirements, identified the technical characteristics considered to be most important for the product:

- Length of pencil
- Time between sharpening
- Lead dust generated
- Hexagonality
- Minimal erasure residue

To obtain the level of technical importance for each of these characteristics, we must quantify the impact of each individual customer requirement. By applying the relationships in Equations (4.1) and (4.2), we obtain the results illustrated in Figure 4.10. For example, the level of importance of the characteristic "length of pencil" is calculated as follows:

$$w_1 = 17 \cdot 3 + 42 \cdot 1 + 17 \cdot 1 = 51 + 41 + 17 = 109$$

whereas the absolute weight of the same characteristic is given by:

$$W_1 = 11.3 \cdot 3 + 53.1 \cdot 1 + 15.1 \cdot 1 \cong 102.27$$

The method of *independent scoring*, traditionally utilized in QFD, presents some inconveniences in the implementation. Some examples include effecting an "arbitrary" conversion of the symbols in the relationship matrix, or tying the results of the prioritization to the "total number of characteristics" of the product under discussion. As we shall see later, these difficulties can in part be overcome by using the coefficient normalization in the relationship matrix, or by utilizing alternative prioritization techniques. These and other aspects will undergo further detailed investigation in Chapters 7 and 8.

4.6.2 Technical Benchmarking

The HoQ, besides the aspects that have been evidenced so far, enables us to *test technical competitiveness*. The numerical value of each characteristic is compared with the appropriate reference value adopted by competitors to indicate the level of competitiveness. This entails perfecting evaluation tests and gathering as much relevant information as possible from all available sources, for example, from technical catalogs. The measurements taken directly by the company on its own product and on competitive products ensure that the comparisons are made under identical conditions using the same equipment and methods.

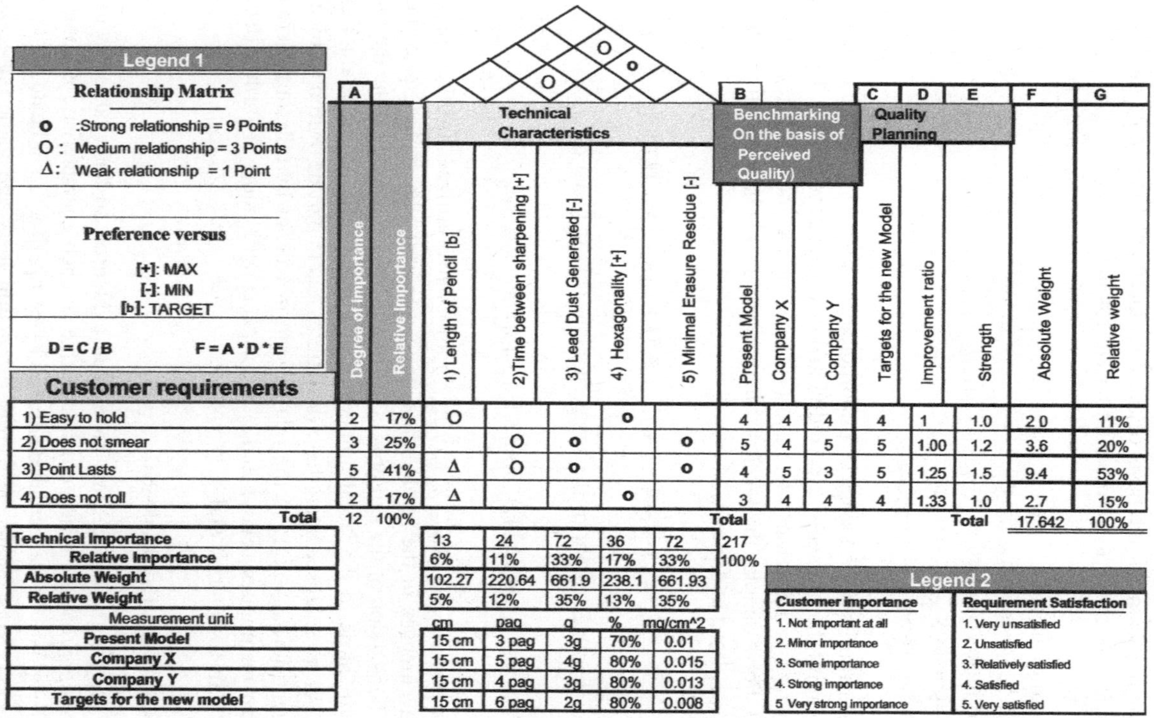

FIGURE 4.10 The house of quality for the planning of a pencil. (From Wasserman, G.S. [1993], *IIE Trans.*, 25(3), 59–65. With permission.)

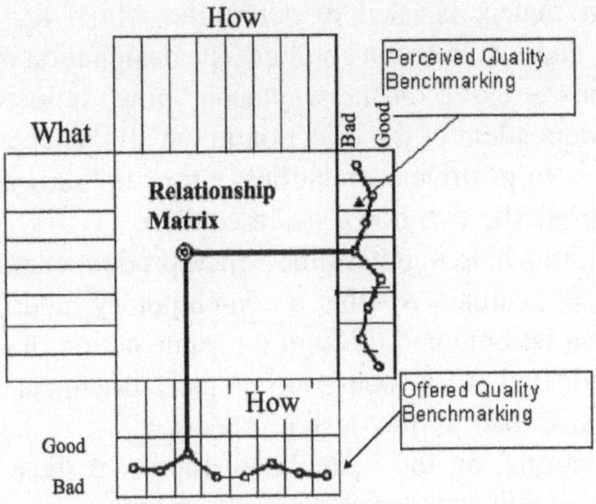

FIGURE 4.11 Competitiveness analysis and benchmarking. QFD represents an ideal environment for quantifying the concepts of offered quality and perceived quality. (From Eureka, W.E. and Ryan, N.E. [1988], *The Customer-Driven Company*, ASI Press, Dearborn, MI; Franceschini, F. and Rossetto, S. [1995], *Res. Eng. Design*, 7, 270–278. With permission.)

The results obtained from the benchmarking analysis on the values of the product technical characteristics are extremely useful because they also allow us to discover slight differences in opinions among technical staff. They enable us to underline those cases where the internal company evaluations do not coincide with the evaluations given by the customer during the phase of *benchmarking on the basis of perceived quality* (Figure 4.11).

4.6.3 Determining Target Values

For each product characteristic, we determine the target values to be used as input data in a design, according to their importance and to the benchmarking analysis.

In the example of the pencil (see Figure 4.10), the characteristics that were evaluated to be most important were "lead dust generated" and the "minimal erasure residue." The new model the company plans must have improved performance values on these points, compared with competitive products.

Another more articulated method used to establish the target settings in planning a new product is illustrated and discussed in Chapter 6.

4.7 CORRELATIONS AMONG CHARACTERISTICS

The correlation matrix placed over the "roof" of the HoQ (see area 6 of Figure 4.1) is triangular in shape and is situated over the area of product characteristics. This matrix allows us to describe the correlation existing among the various technical characteristics, through the use of univocal qualitative symbols representing the positive or negative trend and the intensity of each correlation. To represent these correlations the symbols indicated in Figure 4.6 are used. By highlighting the conflicting relationships (negative or highly negative), the matrix makes for speedy solutions and equitable trade-offs.

The correlation matrix is used to determine which technical characteristics uphold one another and which are in conflict. The designation of positive or negative trends in correlations is based on the way each "how" influences the achievement of other "hows," independent of the direction in which the target setting of the given characteristic moves. In positive correlations, a how upholds another how, whereas in negative correlations the two hows are in conflict.

Positive correlations help to determine which product characteristics are closely related. Thus, we can evaluate whether a contemporary modification of more than one specification can be obtained through the same action on the overall plan, and avoid possible duplicated work loads on company organizational structure, minimizing the energy absorbed by the design process.

Negative correlations, on the other hand, represent those situations that may probably require equitable trade-offs: *these are the situations that should never be ignored.* The unidentified compromises as well as the unresolved compromises inevitably lead to customer requisites not being satisfied. Compromises must be reached through adjustments in target values of the technical characteristics, which represent the system we intend to design.

In the example illustrated in Figure 4.10, the two characteristics of "lead dust generated" and "minimal erasure residue" are strongly related. When trying to produce a pencil that satisfies customer requirements by generating a minimal amount of lead dust, the residual traces on a sheet of paper after erasure will also be diminished.

REFERENCES

Akao, Y. (1988), *Quality Function Deployment*, Productivity Press, Cambridge, MA.

Armacost, R.L., Componation, P.J., Mullens, M.A, and Swart, W.W. (1994), An AHP framework for prioritizing customer requirement in QFD: an industrialized housing application, *IIE Trans.*, 26(4), 72–79.

Brunswik, E. (1952), *The Conceptual Framework of Psychology*, University of Chicago Press, Chicago.

Conti, T. (1992), *Come costruire la Qualità Totale*, Ed. Sperling & Kupfer, Milano.

Dahlgaard, C., Kristensen, D., and Kanji, G. (1994), Break down barriers between departments, in *Advances in Total Quality Management*, Kanji, G., Ed., Carfax, Sheffield, pp. 81–89.

Eureka, W.E. and Ryan, N.E. (1988), *The Customer-Driven Company*, ASI Press, Dearborn, MI.

Franceschini, F. and Rossetto, S. (1995), QFD: the problem of comparing technical/engineering design requirements, *Res. Eng. Design*, 7, 270–278.

Griffin, A.J. and Hauser, J.R. (1993), Patterns of communication among marketing engineering and manufacturing — a comparison between two new product teams, *Manage. Sci.*, 30(3), 360–373.

Hauser, J.R. and Clausing, D. (1988), The House of Quality, *Harv. Bus. Rev.*, 66(3), 63–73.

Hunter, M.R. and Van Landingham, R.D. (1994), Listening to the customer using QFD, *Qual. Prog.*, 27, 56.

Kano, N., Seraku, N., Takahashi, F., and Tsuji, S. (1984), Attractive quality and must-be quality, *J. Jpn. Soc. Qual. Control*, 14(2), 39–48.

Leoni, G. and Raimondi, M. (1993), Un metodo per la misurazione della "customer satisfaction", in *Customer Satisfaction. Misurare e gestire la soddisfazione del Cliente,* GRAMMA, Ed. UTET, Torino.

Sanchez, S.M., Ramberg, J.S., Fiero, J., and Pignatello, J.J. (1993), Quality by Design, in *Concurrent Engineering*, Kusiak, A., Ed., John Wiley & Sons, pp. 235–250.

Tosalli, A., Conti, T., Pettigiani, A., and Pettigiani, M.G. (1990), *La Qualità nel servizio,* Bariletti Editori, Roma, pp. 209–214.

Urban, G.L. and Hauser, J.R. (1993), *Design and Marketing of New Products,* Prentice Hall, Englewood Cliffs, NJ, pp. 267–279.

Wasserman, G.S. (1993), On how to prioritize design requirements during the QFD planning process, *IIE Trans.,* 25(3), 59–65.

Wolfe, M. (1994), Development of the city of quality: a hypertext-based group decision support system for quality function deployment, *Decision Support Syst.,* 11, 303.

5 Supporting Tools of Quality Function Deployment

5.1 INTRODUCTION

Although many maintain that quality function deployment (QFD) methodology is a very useful communications tool, others point out that the technique, in its traditional form, does not come to terms in a sufficiently rigorous manner with some problems that arise when we endeavor to apply it within a complex industrial context.

This effectively represents one of the QFD shortcomings when it is applied in its more traditional form. It often appears to be somewhat "coarse" in its tentative efforts to reach a swift and simple solution to problems that are generally rather complex.

In this chapter we intend to point out some of these weak points found in traditional QFD methodology, and to indicate possible ways of solving them through its integration with several other design-supporting techniques.

Nowadays, QFD methodology is considered to be useful, particularly for its benefits in planning. In the very near future it could come to constitute the cohesive element within a group of instruments able to create an integrated environment for decisional aids in the field of design.

5.2 ASSIGNING LEVELS OF IMPORTANCE TO CUSTOMER REQUIREMENTS

The first aspect we intend to deal with concerns the assignation of numerical values (or *weights*) to the levels of importance of the various needs expressed by the customer. This problem, as we have seen in Chapter 4, Sections 4.2.5 and 4.5.2, has been solved in the traditional QFD version in a rather Spartan manner by applying methods of direct attribution or similar means. On the other hand, we must remember that assigning relative importance values to evaluation criteria constitutes a classic problem in the field of decision-aiding methods.

From this point of view QFD may be considered as a tool aiding designers' decisions when it is utilized to determine an evaluation of importance among various potential processes involved in planning (product or service characteristics).

Customers establish the criteria on which such an evaluation is based. These criteria are the very needs (expressed or latent) connected to that particular product or service. The *decision makers* in this process are, therefore, the customers themselves, and QFD is a means of keeping their preferences in mind during the planning process. The information obtained from the weights assigned to the various requirements–criteria serves, within this context, to aggregate the preferences of customers–decision makers and must reflect their system of values.

The problem of determining the value to be assigned to the weight of each need is, therefore, particularly important because it causes QFD to make the jump from its standing as a purely organizational instrument to the rank of a decisional supporting tool. Among the various methods used to determine the level of importance to be attributed to customer requirements, as seen in Chapter 4, we undoubtedly encounter the analytical hierarchy process (AHP) method [Saaty, 1990a].

5.2.1 General Principles of the Analytical Hierarchy Process Method

The AHP is a technique aiding decision makers, perfected by Saaty during the 1970s. The methodology is particularly useful for the evaluation of complex alternatives having a multiplicity of decisional attributes, where *subjective* criteria are also involved.

The axiomatic fundamentals of the AHP method are described by Saaty [1986]. Essentially, the method organizes the decision into three distinct operative phases:

1. The breakdown of the initial decisional problem into a number of subproblems, which may be more easily understood and evaluated through the construction of a hierarchy of criteria, subcriteria, alternatives, etc., on which the decision is based
2. The determination of a scale of priorities for each level of the decisional hierarchy against which to compare the various elements
3. The evaluation of the consistency of evaluations formerly expressed

5.2.1.1 Hierarchy of Attributes

The construction of hierarchies is one of the basic elements in the process of human comprehension. A decisional problem can be functionally broken down into a functional hierarchy made up of criteria, subcriteria, alternatives, etc., emphasizing their fundamental interactions. The highest level of the hierarchy is univocal (this is *the focus*, or *goal*), whereas the other levels usually comprise several elements. There is no limit set to the number of levels in the hierarchy: when two levels of the hierarchy cannot be directly compared with one another, then another more detailed level has to be created. The hierarchies are flexible in the sense that they can be updated to take into account possible additional criteria.

5.2.1.2 Priorities among Attributes

Once a hierarchy has been established, it is necessary to determine the priorities among elements standing at the same hierarchical level, verifying the consistency of the expressed opinions.

In the AHP method, the priorities are calculated according to the assertions made by a decision maker, who is requested to compare all the elements found on a given decisional level, two at a time. The estimation of their importance (relative preference) is made in relation to the element on the level directly preceding it (above it) in the hierarchy. For example, individual decision makers are asked to judge how important attribute A is compared with attribute B (where A and B are attributes both belonging to the same level in the hierarchy) to guarantee attainment of the objective indicated by the successive element on the next higher level.

If a given level in the hierarchy includes n elements for comparison, $n(n-1)/2$ comparisons of pairs of elements will have to be made. The comparisons are carried out thinking of evaluations expressed on a *ratio scale*. The very fact that it is possible to compare pairs on the basis of ratio scales constitutes one of the axioms of this method.

The evaluation scale $E = (1 \text{ to } 9)$ used by Saaty finds application in many other operative contexts and is rooted in the field of cognitive psychology. Experiments have proved that Saaty's scale of nine elements captures fairly well the preferences of an individual. However, the scale can be altered to suit any individual's needs [Harker and Vargas, 1987].

This scale permits us to express options of preference between two objects in terms of: equal importance (1), moderate importance of one over another (3), essential or strong importance (5), very strong importance (7), extreme importance (9). The values 2, 4, 6, and 8 express the intermediate values between two adjacent judgments.

The operative steps that must be taken to establish priorities are as follows:

1. To effect comparisons among all the various elements found on the same hierarchical level
2. To construct a matrix showing comparisons among pairs
3. To show in matrix **A** the comparisons of the pairs by using the numbers in the scale from 1 to 9
4. To set the elements of the matrix diagonal at 1
5. To show that if k is the value associated to the comparison between the i-th and the j-th alternatives (the coefficient a_{ij} of matrix **A**), then it follows that $a_{ji} = 1/k$ (matrix **A** is a reciprocal matrix)

5.2.1.3 Synthesis of Priorities

Once we have determined the element priorities standing on the same hierarchical level, it becomes important to synthesize the priorities expressed on all levels.

The priorities on each level are weighed against the priority of the attribute on the next higher level used to effect the comparison. Every comparison between a pair of entities (alternatives, subcriteria, or criteria) represents an estimate of the relationship between the priorities or the weights of the two elements that are compared. By applying Saaty's method to these data, we are able to calculate the evaluations of their weights (priorities) on each level of the hierarchy.

5.2.2 Intuitive Justification of the Method for Calculating Weights

A rigorous mathematical justification of the AHP method starting from the definition of the axioms on which it is based will not be given in this book; however, it can be found in various texts shown in the references [Saaty, 1990a]. To understand this method we do, however, present an intuitive justification [Saaty, 1990b].

Let $C_1, C_2, \ldots, C_n$ be a set of n objects (alternatives or criteria) of the same level on a hierarchy. The judgments arising from a comparative evaluation of the single objects are given on an n by n matrix:

$$\mathbf{A} = \{a_{ij}\} \qquad (i, j = 1, 2, \ldots, n)$$

The coefficients of the matrix **A** are defined according to the following rules:

1. If $a_{ij} = \alpha$, then $a_{ji} = 1/\alpha$, with $\alpha \neq 0$, and possibly $\alpha \in E = (1 \text{ to } 9)$.
2. If C_i is as important as C_j, then $a_{ij} = 1$, $a_{ji} = 1$; in particular, $a_{ii} = 1$ $\forall i = 1, 2, \ldots, n$.

Thus, for matrix $\mathbf{A} \in R^{n,n}$ (reciprocal) of the comparisons on pairs:

$$\mathbf{A} = \begin{bmatrix} 1 & a_{12} & \cdots & a_{1n} \\ 1/a_{12} & 1 & \cdots & a_{2n} \\ \cdots & \cdots & \cdots & \cdots \\ 1/a_{1n} & 1/a_{2n} & \cdots & 1 \end{bmatrix}$$

the problem lies in the assigning to the n objects $C_1, C_2, \ldots, C_n$, a set of numerical weights $w_1, w_2, \ldots, w_n$, which reflect the judgments expressed in **A**.

If, for example, the preferences expressed were the result of typically physical measurements (length, mass, etc.), the weight would express the ratio between the determined values:

$$w_i/w_j = a_{ij} \qquad (\forall i, j = 1, 2, \ldots, n)$$

With this position matrix **A** becomes:

$$\mathbf{A} = \begin{bmatrix} w_1/w_1 & w_1/w_2 & \cdots & w_1/w_n \\ w_2/w_1 & w_2/w_2 & \cdots & w_2/w_n \\ \vdots & \vdots & \ddots & \vdots \\ w_n/w_1 & w_n/w_2 & \cdots & w_n/w_n \end{bmatrix}$$

In this case the coefficients of matrix **A** satisfy the following properties:

$$(*)\quad a_{ij}\cdot a_{jk}=\frac{w_i}{w_j}\cdot\frac{w_j}{w_k}=\frac{w_i}{w_k}=a_{ik}\quad \text{(Consistency)}$$

$$(**)\quad a_{ji}=\frac{w_j}{w_i}=\frac{1}{w_i/w_j}=\frac{1}{a_{ij}}\quad \text{(Reciprocity)}$$

Let us now consider the resolution of a linear system:

$$\mathrm{A}\cdot\mathbf{x}=\mathbf{y} \tag{5.1}$$

with

$$\mathbf{x}=\begin{bmatrix}x_1\\ \vdots\\ x_n\end{bmatrix}$$

$$\mathbf{y}=\begin{bmatrix}y_1\\ \vdots\\ y_n\end{bmatrix}$$

By making Equation (5.1) explicit we find:

$$\sum_{j=1}^{n} a_{ij}\cdot x_j=y_j$$

$$i=1,\ \ldots,\ n$$

On the other hand, the equation $a_{ij}=\frac{w_i}{w_j}$ can be rewritten, by multiplying both terms by $\frac{w_j}{w_i}$ in the following manner:

$$a_{ij}\cdot\frac{w_j}{w_i}=1$$

$$(\forall i,\ j=1,\ \ldots,\ n)$$

By summing with respect to j we obtain:

$$\sum_{j=1}^{n} a_{ij} \cdot \frac{w_j}{w_i} = n$$

$$i = 1, \ldots, n$$

or

$$\sum_{j=1}^{n} a_{ij} \cdot w_j = n \cdot w_i$$

$$i = 1, \ldots, n$$

which is equivalent to:

$$\mathrm{A} \cdot \mathbf{w} = n \cdot \mathbf{w} \tag{5.2}$$

This expression shows the concept that **w** (vector of the weights that qualifies each single alternative) is an *eigenvector* of **A**, with n one of its *eigenvalues*.

$$\begin{bmatrix} 1 & a_{12} & \cdots & a_{1n} \\ 1/a_{12} & 1 & \cdots & a_{2n} \\ \vdots & \vdots & \ddots & \vdots \\ 1/a_{1n} & 1/a_{2n} & \cdots & 1 \end{bmatrix} \cdot \begin{bmatrix} w_1 \\ \vdots \\ \vdots \\ w_n \end{bmatrix} = n \cdot \begin{bmatrix} w_1 \\ \vdots \\ \vdots \\ w_n \end{bmatrix}$$

If we now analyze a practical example where the coefficient a_{ij} expresses an *intensity of preference* between two alternatives not based on exact (physical) measures, but on subjective judgments, normally a_{ij} will derive from the ideal relationship $\frac{w_i}{w_j}$ and the relationship A·**w** = n·**w** will no longer hold.

At this point two important concepts in the matrix theory intervene:

1. If $\lambda_1, \ldots, \lambda_n$ are the n solutions (eigenvalues) of the linear system:

$$\mathrm{A} \cdot \mathbf{x} = \lambda \cdot \mathbf{x} \tag{5.2a}$$

$$\begin{cases} \mathbf{A} \in \Re^{n,n} \\ \mathbf{x} \in \Re^{n,1} - \{\mathbf{0}\} \\ \lambda \in \Re \end{cases}$$

and if $a_{ii} = 1$; $\forall i = 1, 2, \ldots, n$, then:

$$\sum_{i=1}^{n} \lambda_i = n \tag{5.3}$$

If Equations (5.2) and (5.3) are true, then all the eigenvalues of **A** are null, except the greatest whose value is n (because **A** is reciprocal).

2. If we vary slightly the coefficients a_{ij} of a reciprocal matrix **A**, the auto-values will also vary slightly.

In combining the results in 1 and 2 by the very fact that our matrix **A** has only unit values on its principal diagonal $a_{ii} = 1$; $\forall i = 1, 2, \ldots, n$, and if **A** is consistent (i.e., if $a_{ik} = a_{ij} \cdot a_{jk}$), then small variations of the a_{ij} keep the largest eigenvalue λ_{max} close to n, whereas the other eigenvalues will remain close to zero.

Once the matrix of pairwise comparisons **A** has been assigned with the objective of defining the vector of priorities, it is simply a case of determining the eigenvector **w** associated to the eigenvalue λ_{max} that satisfies the equation:

$$\mathbf{A} \cdot \mathbf{w} = \lambda_{max} \cdot \mathbf{w} \tag{5.4}$$

Once $\mathbf{w} = \begin{bmatrix} w_1 \\ \vdots \\ w_n \end{bmatrix}$ has been determined, the weights are successively normalized, for easier interpretation:

$$w_i' = \frac{w_i}{\sum_{i=1}^{n} w_i} \quad \text{with} \sum_{i=1}^{n} w_i' = 1 \qquad \forall i = 1, 2, \ldots, n \tag{5.5}$$

The weights w_i' represent the relative importance of the entities that have been compared.

5.2.2.1 Consistency Evaluation

An important consideration to be made when utilizing the AHP method is the notion of consistency. Customers who answer that a certain characteristic A is twice as important as characteristic B, and that B is three times more important than D, will be supplying a consistent evaluation if they also answer that the characteristic A is six times more important than characteristic D. Any other value assigned by customers when comparing the characteristics A to D will render their opinions not

TABLE 5.1
Average RI Values

N	1	2	3	4	5	6	7	8
RI	0	0	0.58	0.90	1.12	1.24	1.32	1.41

From Saaty, T.L. (1990a), *Multicriteria Decision Making: The Analytical Hierarchy Process*, 2nd ed., RWS Publications, Pittsburgh. With permission.)

coherent or not consistent. The method involving eigenvalues allows us to evaluate quantitatively the distance from condition of consistency. As small variations in a_{ij} imply small variations in λ_{max}, the difference $(\lambda_{max} - n)$ can be taken to be a measure of consistency of the evaluations expressed in matrix **A**.

We define the *consistency index* as the ratio:

$$CI = \frac{\lambda_{max} - n}{n - 1} \tag{5.6}$$

CI is compared with the *random index* (RI) randomly generated for reciprocal matrices, with reciprocals forced, having n varying from 1 to 15 and taking into account the average on a sample having an increasing number of units (from 100 to 500) (Table 5.1).

The ratio:

$$CR = \frac{CI}{RI} \tag{5.7}$$

defines the so-called consistency ratio (CR).

An empirical rule supplied by Saaty states that the CR of 0.10 or less is considered acceptable. When judgments are not far too coherent, decision makers should be given the opportunity to have another look at their pair comparisons.

5.2.3 Advantages and Disadvantages of Integrating Quality Function Deployment and Analytical Hierarchy Process

Many authors have suggested the use of the AHP method in applying QFD [Akao, 1990; Aswad, 1989]. AHP is used to assign a level of priorities to a hierarchy of requirements that constitute the customer's own criteria of evaluation. The hierarchy is that found in the tree or table of demanded quality. The customer's basic requirements constitute the strategic dimensions of the decisional problem, the second level includes the criteria of evaluation, and the lowest level includes the attributes or the subattributes.

The advantage of using AHP to determine the priorities of customer requirements is that it constitutes a *ready-made* model enabling us to deal with complex situations of order ranking, based on subjective criteria.

The calculation of the various weights is facilitated, for example, by the use of the computer program Expert Choice™, which implements Saaty's procedure. A considerable advantage of the method is that it supplies a measurement of the consistency of evaluations expressed by the customer–decision maker. Moreover, it points out which judgments are the more incoherent among those expressed by the customer, permitting reevaluations where necessary.

In this sense, if the number of customers–decision makers is limited and if they can be grouped, AHP can be used in a quite advantageous manner as a *group decision support system* (GDSS) making use of a *moderator* whose duty it will be to facilitate reaching a consensus on the evaluations expressed by the various customers.

The situation changes remarkably when dealing with a rather numerous but scattered group of customers, as may very often happen where some particular industrial products are involved. In any case, even here, the AHP method can be used.

Aczel and Saaty [1983] have in fact demonstrated that matrix **A′** of pairwise comparisons obtained by calculating the geometric average of the *judgments expressed by n* customers:

$$A' = \left\{a'_{ij}\right\}$$

with

$$a'_{ij} = \sqrt[n]{a_{ij}^{[1]} \cdot a_{ij}^{[2]} \cdot \ldots \cdot a_{ij}^{[k]} \cdot \ldots \cdot a_{ij}^{[n]}} \tag{5.8}$$

(where $a_{ij}^{[k]}$ is the evaluation of the k-th customer on the comparison between the pair of requisites i and j), maintains the property of reciprocity and its elements still belong to the ratio scale E = (1 to 9). It is, therefore, possible to apply the method of eigenvalues to matrix **A′** that synthesizes the evaluations of n customers.

Nonetheless, if the customers are numerous and cannot be easily contacted, we shall have to utilize a questionnaire (the program Expert Choice™ creates one automatically), which is rather difficult to compile and much more irksome for the customer. It requires a greater number of evaluations than does a traditional questionnaire, which does not require pair comparisons of requirements on the same level.

Furthermore, it must be pointed out that *the customer finds it extremely difficult to indicate precisely how many times* one alternative may be more important than another. Does the information supplied by customers have the properties of a ratio scale?

The danger of noncoherence between judgments then becomes quite evident, as the difficulty of adjusting them is equally evident. We could run the risk, too, of possessing a heap of data of somewhat scant significance and using them *uncritically or improperly.*

●: 9 points O : 3 points Δ : 1 point		Technical Characteristics										
		S 1									S2	
Customer Needs	Degree of Importance	1.1	1.2	1.3	1.4	1.5	1.6	1.7	1.8	1.9	2.1	2.2
R1	10%	●	●	●	●	●	●	●	●	●		
R2	90%										●	●
		90	90	90	90	90	90	90	90	90	810	810
	Absolute Importance	810									1620	
	Relative Importance	33.3%									66.6%	

FIGURE 5.1 The need for deployment normalization: an exaggerated example. (From Wasserman, G.S. [1993], *IIE Trans.*, 25(3), 59–65. With permission.)

5.3 PRIORITIZING THE TECHNICAL CHARACTERISTICS

We have seen in Chapter 4, Section 4.6 how it may be possible through QFD to determine the prioritization of technical characteristics. The information contained in the relationship matrix is used to draw up a list ranking the *level of attention* a designer will have to dedicate to each single product characteristic (keeping in mind the relative importance of customer requirements).

It is, therefore, of fundamental importance that customer requirements be considered without forcing or distorting the customer's intentions. In this respect, the traditional method described in Chapter 4 does present some drawbacks. In this chapter as well as the one following, we will deal with how the level of importance of a characteristic depends on the number of subcharacteristics used to describe it in detail. Such problems can be opportunely resolved by *normalizing* the relationship matrix.

Let us consider a case that has been purposely exaggerated. We have, for example, two customer requirements: *R*1, a requisite of relatively *little importance* (weighing 10%), and *R*2, a requisite relatively *important* (weighing 90%) (Figure 5.1).

We also have two design characteristics, *S*1 and *S*2, which completely characterize the product to be planned in terms of satisfaction of customer requirements. Note that *S*1 is a level one design characteristic that declines into no less than 9 subcharacteristics on a lower level, coded 1.1 to 1.9, respectively.

Each one of these subcharacteristics is strongly correlated to the requirement having little importance, *R*1. On the other hand, specification *S*2 declines into two subcharacteristics (2.1 and 2.2), each of which is strongly correlated to the important requirement *R*2.

If we apply the traditional *independent scoring method* [Akao, 1990] with its scale of 1 to 3 to 9, we will determine that the weight of the characteristic $S1$, connected to requisite $R1$ (and none other), is 33.3%, whereas the weight of characteristic $S2$, able to satisfy requirement $R2$, is 66.7%.

According to the levels of importance of each customer requirement, the contribution of the requisites $R1$ and $R2$ on an overall level of customer satisfaction is 10 and 90%, respectively. We should expect a similar proportion (1:9) between the levels of relative importance of the planning characteristics $S1$ and $S2$. On the contrary, we note that because the first characteristic is detailed by a greater number of subcharacteristics than the second characteristic, the level of relative importance is *artificially* heightened from 10 to 33.3%.

5.4 NORMALIZING THE COEFFICIENTS OF THE RELATIONSHIP MATRIX

5.4.1 Lyman's Normalization

To solve this problem, Lyman [1990] proposes the normalization of the coefficients in the relationship matrix **R**. The coefficients $\tilde{r}_{ij}$ of relationship matrix $\tilde{\mathbf{R}}$ are obtained by dividing each of the coefficients of **R** by the sum of the values on each line.

We will therefore obtain:

$$\tilde{r}_{ij} = \frac{r_{ij}}{\sum_{j=1}^{m} r_{ij}} \tag{5.9}$$

In this manner, the resultant matrix $\tilde{\mathbf{R}}$ will satisfy the property that the sum of the elements on each row is equal to 1. This property can be expressed in matrix form in the following manner:

$$\tilde{\mathbf{R}} \cdot \mathbf{1} = \mathbf{1} \tag{5.10}$$

as $\mathbf{1} = (1, 1, 1, \ldots, 1)^T$.

Figure 5.2 illustrates the effect of Lyman's normalization on the example shown in Figure 5.1. Normalization contributes to ensuring that the calculated weights of the design characteristics truly reflect the ranking order of customer requirements on the basis of their importance. This procedure does not take into account the fact that the technical characteristics of a product may be correlated to one another. In practice this happens very frequently. Sometimes it is possible to reduce the level of dependency between the design characteristics by eliminating those that appear to be redundant, or by devising planning alternatives to reduce correlation.

		Technical Characteristics											
		S 1									S2		
Customer Needs	Degree of importance	1.1	1.2	1.3	1.4	1.5	1.6	1.7	1.8	1.9	2.1	2.2	Row sum
R1	10%	o	o	o	o	o	o	o	o	o			81
R2	90%										o	o	18
		1.11	1.11	1.11	1.11	1.11	1.11	1.11	1.11	1.11	45	45	
	Absolute importance	10									90		
	Relative importance	10%									90%		

FIGURE 5.2 The need for deployment normalization: an exaggerated example. (From Lyman, D. [1990], Deployment Normalization, 2nd Symposium on QFD cosponsored by ASCQ and ASI, pp. 307–315; Wasserman, G.S. [1993], *IIE Trans.*, 25(3), 59–65. With permission.)

5.4.2 Wasserman's Normalization

An extension of Lyman's normalization procedure has been suggested by Wasserman [1993] to solve the problem of interdependant planning characteristics. To model the dependency, we determine the vector space of technical characteristics and that of customer requirements. Let us assume that the vector space of customer requisites $\mathfrak{I}$ is generated by the unit vectors $\{\mathbf{u}_i\}$, $i = 1, 2, \ldots, n$. For the moment let us assume that the customer requirements are not correlated. Thus, the set of vectors $\{\mathbf{u}_i\}$, $i = 1, 2, \ldots, n$ forms an orthonormal basis spanning the customer requirements space $\mathfrak{I}$. This hypothesis is coherent with the way the customer requirements are usually treated during the QFD process.

If necessary, the interdependence of the customer requirements may be analyzed later. At this point we are able to write the vector **d** of customer importance rating in the following manner:

$$\mathbf{d} = d_1 \cdot \mathbf{u}_1 + d_2 \cdot \mathbf{u}_2 + \ldots + d_n \cdot \mathbf{u}_n \tag{5.11}$$

where d_i is the importance of the i-th requirement.

On the other hand, to model the vector space Ξ of product characteristics, let us assume that it is generated by the unit vectors $\{\mathbf{v}_j\}$ that do not necessarily constitute an orthonormal basis for Ξ, because they may be linearly dependent on one another. To represent the interdependence of the technical characteristics, we will introduce the notation γ_{jk} to indicate the element on the roof of the house of quality (HoQ) that describes the intensity of correlation existing between characteristic j and characteristic k. We can note this as:

$$\gamma_{jk} \equiv \mathbf{v}_j \cdot \mathbf{v}_k \left(\equiv \cos\left(\mathbf{v}_j, \mathbf{v}_k\right)\right) \tag{5.12}$$

obviously $\gamma_{jj} \equiv \mathbf{v}_j \cdot \mathbf{v}_j = 1, \ \forall j = 1, \ \ldots, \ m$

Coefficients γ_{jk} expressing the intensity of correlation between characteristics are assigned by the designers using a scale very similar to that used for the coefficients of the relationship matrix. However, it is necessary to make sure the scale levels remain between 0 and 1. Thus, a value of 0.9 will be assigned to a strong correlation, 0.3 to a medium correlation, and 0.1 to a weak correlation. In the example of the pencil (see Figure 4.10), the vectors $\{\mathbf{v}_j\}$ are

$$\gamma_{12} = \gamma_{21} = 0$$

$$\gamma_{13} = \gamma_{31} = 0$$

$$\gamma_{14} = \gamma_{41} = 0$$

$$\gamma_{15} = \gamma_{51} = 0$$

$$\gamma_{23} = \gamma_{32} = 0$$

$$\gamma_{24} = \gamma_{42} = 0$$

$$\gamma_{25} = \gamma_{52} = 0$$

$$\gamma_{34} = \gamma_{43} = 0$$

$$\gamma_{35} = \gamma_{53} = 0$$

$$\gamma_{45} = \gamma_{54} = 0$$

A generalization of Lyman's normalization referred to correlated planning characteristics is, therefore, as follows:

$$\left(r_{i,j}^{\text{norm}} \cdot \mathbf{v}_1 + r_{i,2}^{\text{norm}} \cdot \mathbf{v}_2 + \ldots + r_{i,n}^{\text{norm}} \cdot \mathbf{v}_n\right) \cdot \left(\mathbf{v}_1 + \mathbf{v}_2 + \ldots + \mathbf{v}_n\right) = 1$$

with

$$i = 1, \ 2, \ \ldots, \ n$$

which, as we have:

$$\left\{\sum_{i=1}^{n} \mathbf{v}_i\right\} \cdot \left\{\sum_{j=1}^{n} \mathbf{v}_j\right\} = \left\{\sum_{i=1}^{n} \sum_{j=1}^{n} \gamma_{ij}\right\}$$

is satisfied by calculating the normalized coefficients using the following formula:

$$r_{i,j}^{\text{norm}} = \frac{\sum_{k=1}^{m}\left(r_{i,k}\cdot\gamma_{k,j}\right)}{\sum_{j=1}^{m}\sum_{k=1}^{m}\left(r_{i,j}\cdot\gamma_{j,k}\right)}$$

Notice that this type of transformation is referable to Lyman's normalization where the design characteristics are independent from one to the other, such that:

$$\gamma_{jk} = 1 \quad \text{if } j = k$$

otherwise

$$\gamma_{jk} = 0$$

For example, the relation coefficient $r_{3,1}^{\text{norm}}$ in the example of the pencil is calculated in the following manner:

$$r_{3,1}^{\text{norm}} = \frac{\sum_{k=1}^{5}\left(r_{3,k}\cdot\gamma_{k,1}\right)}{\sum_{j=1}^{5}\sum_{k=1}^{5}\left(r_{3,j}\cdot\gamma_{j,k}\right)}$$

$$= \frac{r_{3,1}\cdot\gamma_{1,1}+r_{3,2}\cdot\gamma_{2,1}+r_{3,3}\cdot\gamma_{3,1}+r_{3,4}\cdot\gamma_{4,1}+r_{3,5}\cdot\gamma_{5,1}}{r_{3,1}\cdot\left(\gamma_{1,1}+\ldots+\gamma_{1,5}\right)+r_{3,2}\cdot\left(\gamma_{2,1}+\ldots+\gamma_{2,5}\right)+r_{3,3}\cdot\left(\gamma_{3,1}+\ldots+\gamma_{3,5}\right)+r_{3,4}\cdot\left(\gamma_{4,1}+\ldots+\gamma_{4,5}\right)+r_{3,5}\cdot\left(\gamma_{5,1}+\ldots+\gamma_{5,5}\right)}$$

$$= \frac{1\cdot1+3\cdot0+9\cdot0+0\cdot0+9\cdot0}{1\cdot(1)+3\cdot(1.6)+9\cdot(2.2)+0\cdot(1)+9\cdot(2.2)} = \frac{1}{45.4} \cong 0.022$$

Figure 5.3 shows the example of the pencil after Wasserman's process of normalization. It should also be noted that with this transformation the absolute weights (calculated with normalized coefficients) coincide with the relative weights (in percentages).

Even in this simplified example, the effect of normalization is evident. An analysis of the relationship matrix reveals that "minimal erasure residue" and "lead dust generated" are essentially redundant characteristics. We can see from the roof of the HoQ that both characteristics have an equal impact on the customer requirements "does not smear" and "point lasts." Therefore, it may be sufficient to introduce only one of the two to satisfy both customer requirements.

This consideration is pointed out through the procedure of normalization. Before applying it, both specifications had been assigned a level of relative importance equal to 33% (see Figure 4.10). After it had been applied, the level of importance

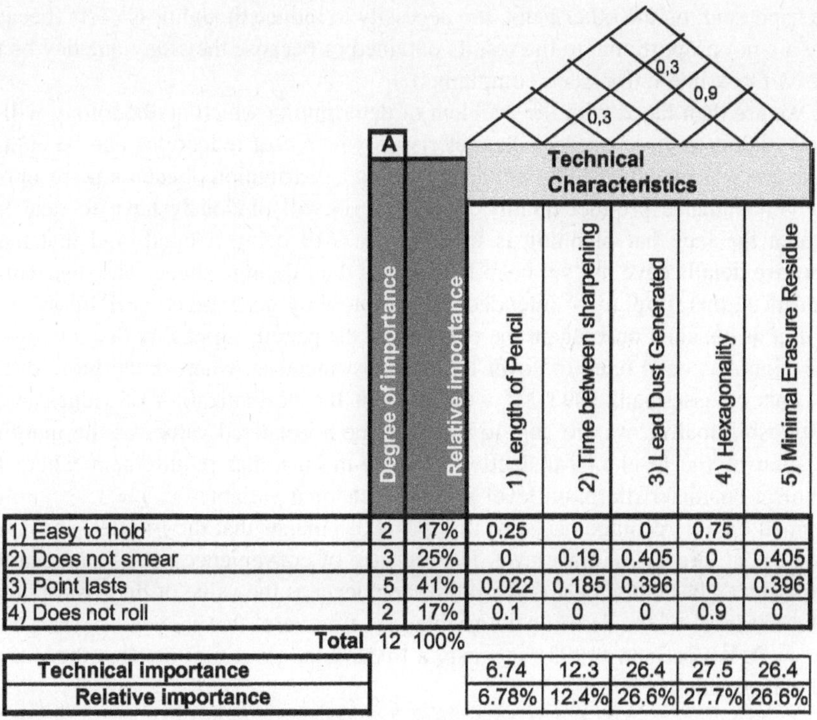

FIGURE 5.3 Wasserman's procedure for the normalization of the pencil example. (From Wasserman, G.S. [1993], *IIE Trans.*, 25(3), 59–65. With permission.)

of each of the two characteristics diminished to 26.7%, with a level of global importance fixed at about 54%. Furthermore, the characteristic "hexagonality," which appears to be a very important characteristic for "easy to hold" and "does not roll" customer requirements, is seen to increase its level of importance after normalization from 17 to 27.5%.

5.5 QUALITY FUNCTION DEPLOYMENT AND VALUE ANALYSIS

5.5.1 Simplified Model for Costing

Numerous authors [King, 1989; Akao, 1990] point out the usefulness of analyzing the costs involved in the QFD planning process.

From the very first stages of planning a new product, companies are obliged to define the market niche and the targeted sales price. Akao talks of a *deployment of costs* to be carried out in parallel with the *deployment* of the quality attributes defined by the customer. The aim of this deployment of costs should be to introduce a systematic procedure for the evaluation and optimization of the cost of the product, without diminishing the importance of its quality and its reliability. This necessity stems from the need to avoid the erroneous allocation of company resources on the one hand, and, on the other hand, the necessity to reduce thoughtless costs (because they are not proportional to the results obtained or because they may one day be the cause of possible unforeseen complaints).

We are then faced with the problem of determining which methodology will be able to indicate those product characteristics where cost reductions can be applied to the greatest advantage. This entails operating a distribution of economic resources that will enhance product quality. The analysis will obviously have to take into account the fact that planning is in the process of being defined, and that many operative details have not yet been frozen into their definite shape. The distribution of costs is, therefore, to be intended as a preliminary generalized attribution.

Let us consider once again the example of the pencil; supposing that a budget B of 2 cents has been fixed to cover the unit cost increase, whereas the basic cost is 10 cents [Wasserman, 1993]. If we carry out the normalization of values in the relationship matrix, we are able to interpret the normalized values as the marginal variation of the level of satisfaction of the j-th customer requirement where the technical characteristic is on level x_j. The decisional variables x_j, $j = 1, \ldots, m$, are assumed on a percentage basis, so that $x_j = 100\%$ means that the j-th design characteristic is at target (optimal) level. For the sake of convenience, let us suppose that the product's basic unit cost is set at the same level as the value of the cost obtained when the level of the decisional variables is set at zero, that is, when $x_1 = x_2 = \cdots = x_m = 0$. Wasserman [1993] suggests a linear cost constraint as described in the following equation:

$$c_1 \cdot x_1 + c_2 \cdot x_2 + \ldots + c_m \cdot x_m \leq B \tag{5.13}$$

The cost coefficients $c_1, c_2, \ldots, c_m$ represent the incremental change in unit cost associated to a variation of x_j.

The decisional variables (design characteristics) are to be so determined with a view to maximizing global CS. The following objective function is proposed:

$$CS = \max\{w_1 \cdot x_1 + w_2 \cdot x_2 + \ldots + w_m \cdot x_m\} \tag{5.14}$$

In this formulation there is an implicit assumption that the level of customer satisfaction will be modified for those technical characteristics whose fulfillment will produce a high marginal increase in the overall *CS*.

The objective function is seen to be a simple, linear weighting of the technical importance measures w_j for the normalized relationship matrix, and the decision variables x_j. Generally, this assumption must be verified according to the particular product under examination, and the capability to render quantifiable in terms of costs all the contributions of the single technical characteristics.

5.5.2 Interpreting the Model

The suggested costing model — Equations (5.13) and (5.14) — takes us back to the classic *knapsack problem* 0-1. Given a series of different-sized items, each one having a specific value, we endeavor to maximize the value of the contents in the knapsack, without overstepping its overall capacity bound B.

It can be shown that an upper bound to the value of contents of the knapsack is obtained by prioritizing the articles, assigning the unit value for each item in decreasing order. If the integer restrictions in the knapsack problem are relaxed, then it can be shown that the optimal packing of the knapsack is obtained by following this ranking exactly.

Although as formerly said, x_j = 100% represents the level of optimal quality status, c_j represents the cost required to reach the target for the j-th characteristic. To neutralize the effects of antagonism among the negatively correlated planning specifications, we assume that the scale on the planning specifications varies in the continuous interval between [–100% and +100%]. In other words, we can allow negative variations in the values of the characteristics starting from their basic values.

Thus, the allocation of resources according to the strategy best suited to solving the knapsack problem should be strictly based on the value of a simple index formed by the (w_j/c_j) ratios, $j = 1, \ldots, m$. This is coherent with the opinions expressed by Hales, Lyman, and Norman [1990], who suggest indicating in QFD matrices the *importance of the effort required*. In effect the *effort required* might be considered, as a first approximation, as a cost proportional to the resources allocated and the organizational efforts required.

5.5.3 Illustrative Example

Returning to the example of the pencil design, let us suppose that the marginal costs connected to reaching an optimal level of performance for the various characteristics are those defined in the column "cost c_j" in Table 5.2. As we can observe, due to the restriction of 2 cents it is impossible to reach an optimal level in all the characteristics, because the necessary expenditure would be of $2.90.

The relative weights w_j (shown in the first column on Table 5.2) tell us that as far as the satisfaction of customer requirements is concerned, the most important characteristics are, in order, 3 and 5 (which are the redundant characteristics), followed by 4, 2, and 1. If we consider the costs, however, characteristic 4 is shown to be the most economical to improve, followed by characteristics 3 and 5, then by 2 and 1 (Table 5.3).

According to the model described in Section 5.5.1 of Chapter 5, to maximize global customer satisfaction while keeping in mind the budgetary restrictions, it is necessary to rank the characteristics on the basis of a weight and cost quotient rating. Successively, the available economic resources will be allocated starting from the technical characteristics having the highest value according to this quotient rating.

Table 5.2 shows the optimal allocation of economic resources resulting in an improvement of the product pencil, keeping to the available budget of 2 cents.

5.6 CONCLUSIONS

In this chapter we have focused our attention on the problems associated with gathering and processing the data essential for working on a product when using

TABLE 5.2
Recommended Allocation of Incremental Design Resources

Technical Characteristic	Technical Importance (%)	Cost c_j (c$)	w_j/c_j (%/c$)	% Allocation	Resources Allocated (c$)
Length of pencil	7	1.00	7	10	0.10
Time between sharpening	12	0.60	20	100	0.60
Lead dust generated	26.5	0.50	53	100	0.50
Hexagonality	28	0.30	93	100	0.30
Minimal erasure residue	26.5	0.50	53	100	0.50
Total	100	2.90			2.00

From Wasserman, G.S. [1993], *IIE Trans.*, 25(3), 59–65. With permission.

TABLE 5.3
Prioritization of Technical Characteristics with Respect to Importance, Cost, and Importance/Cost Index

Technical Characteristic	Prioritization According to Technical Importance	Prioritization According to Cost	Prioritization According to Importance/Cost Index
Length of pencil	5	5	5
Time between sharpening	4	4	4
Lead dust generated	1	2	2
Hexagonality	3	1	1
Minimal erasure residue	1	2	2

From Wasserman, G.S. [1993], *IIE Trans.*, 25(3), 59–65. With permission.

QFD. We have seen that QFD, to be applied, must undergo an attentive phase of design analysis and must include a multitude of carefully gathered information concerning the customer.

Therefore, as we shall notice in Chapter 7, a preeminent role in this direction is played by the use of quantitative and nonquantitative scales, used to gather various responses. Choosing the most suitable scales, how to formulate the questions, and how to gather information* are therefore not problems of secondary importance when using QFD. The choice of one scale instead of another could direct the planning process toward entirely different design solutions.

* According to Urban and Hauser [1993], the design of questionnaires is to be considered an art.

REFERENCES

Aczel, J. and Saaty, T.L. (1983), Procedures for synthesizing ratio judgements, *J. Math. Psychol.,* 27, 93–102.

Akao, Y. (1990), *Quality Function Deployment,* Productivity Press, Cambridge, MA.

Armacost, R.L., Componation, P.J., Mullens, M.A., and Swart, W.W. (1994), An AHP framework for prioritizing customer requirements in QFD: an industrialized housing application, *IIE Trans.,* 26(4), 72–79.

Aswad, A. (1989), Quality Function Deployment: A System Approach, Proceedings of the 1989 IIE Integrated System Conference, Institute of Industrial Engineers, Atlanta, GA, pp. 27–32.

Franceschini, F. and Rossetto, S. (1997), Design for quality: selecting product's technical features, *Qual. Eng.*, 9(4), 681–688.

Fraser, N.M. (1994), Ordinal preference representations, *Theory Decision,* 36(1), 45–67.

Hales, R., Lyman, D., and Norman, R. (1990), Quality Function Deployment and the Expanded House of Quality, Technical Report, International TechneGroup Inc., New York.

Harker, P. and Vargas, LG. (1987), The theory of ratio scale estimation: Saaty's analytic hierarchy process, *Manage. Sci.,* 33(11), 1383–1403.

King, B. (1989), *Better Designs in Half the Time: Implementing QFD in America,* GOAL/QPC, Methuen, MA.

Lyman, D. (1990), Deployment Normalization, 2nd Symposium on QFD cosponsored by ASCQ and ASI, pp. 307–315.

Roy, B. (1990), Decision-aid and Decision-making, *Eur. J. Operational Res.,* 45, 324–331.

Roy, B. (1991), The outranking approach and the foundations of ELECTRE methods, *Theory Decision,* 31(1), 49–73.

Saaty, T.L. (1986), Axiomatic foundation of the analytic hierarchy process, *Manage. Sci.,* 32(7), 841–855.

Saaty, T.L. (1990a), *Multicriteria Decision Making: The Analytic Hierarchy Process*, 2nd ed., RWS Publications, Pittsburgh.

Saaty, T.L. (1990b), *Decision Making for Leaders: The Analytic Hierarchy Process for Decisions in a Complex World,* rev. ed., RWS Publications, Pittsburgh.

Saaty, T.L. (1990c), *Decision Making, Scaling, and Number Crunching,* in *Decision Making for Leaders: The Analytic Hierarchy Process for Decisions in a Complex World,* rev. ed., RWS Publications, Pittsburgh, pp. 269–274.

Urban, G.L. and Hauser, J.R. (1993), *Design and Marketing of New Products,* Prentice Hall, Englewood Cliff, MI.

Vansnick, J.C. (1986), On the problem of weights in multiple criteria decision making (the noncompensatory approach), *Eur. J. Operational Res.*, 24, 288–294.

Wasserman, G.S. (1993), On how to prioritize design requirements during the QFD planning process, *IIE Trans.*, 25(3), 59–65.

Zahedi, F. (1986), The analytic hierarchy process — a survey of the method and its applications, *Interfaces*, 16, 96–108.

6 Selecting Technical Features of a Product

6.1 INTRODUCTION

The quality problem has become more and more central for enterprises. Initially focused on manufacturing, attention is now devoted *in primis* to the design moment, as the phase conditioning of the product performance and its success throughout its life [Akao, 1992; Franceschini, 1993].

Among emerging methodologies to support product design, quality function deployment (QFD) is playing a special role. It constitutes an adequate environment to carry out a comparative analysis of the product performance with competitors in the market (*benchmarking*) [Bemowski, 1991]. According to a definition by Camp [Zairi, 1992], benchmarking is considered

> the continuous process of measuring our products, services, and business practices against the toughest competitors or those companies recognized as industry leaders.

This comparison, therefore, can represent an effective starting point for the technical characteristics definition with reference to the customer or user requirements. Thus, QFD becomes a tool able to quantify *offered* and *perceived* quality [Franceschini and Rossetto, 1995a], allowing the achievement of the key concept of design oriented to the customer.

In this chapter, the problem of selecting the technical features of a product to make it more competitive from the user's global point of view is illustrated [Garvin, 1987; Franceschini and Rossetto, 1997, 1998]. The presentation is organized in such a way as to highlight the different aspects of the problem connected with the different levels of information available throughout the design. A heuristic algorithm has been developed and discussed with the aim to support the designer in this activity. The algorithm, which operates in QFD environment, allows the definition of the technical and engineering product design characteristics values by means of a comparison with competitors' products, and market expectation.

6.2 PROBLEM FORMULATION

The problem of comparing some design solutions on a set of evaluation criteria is a recurring event in practical applications. The inverse problem of creating a better alternative, minimizing efforts and resources, is as frequent. A typical example of a direct decision problem is a product selection carried out by a customer [Roy and

Bertier, 1972], whereas an inverse problem is one approached by a manufacturer to obtain the technically best, with the constraint of minimizing employed resources.

If more decisional criteria are involved in a decision, the concept of the best solution cannot be read in an absolute key. It must be referred to the preference system of the decision maker. The ability of operating a selection among more alternatives (direct decision problem), compared with a set of evaluation criteria, is the central theme of multicriteria decision aiding (MCDA) [Roy, 1991; Vincke, 1992] and multicriteria decision making (MCDM) techniques [Steuer, 1986], interpreted according to their specificity [Roy, 1990]. These methods facilitate the analysis, the modelization, and the synthesis of preferences expressed by a decision maker in the comparison of two alternatives a and a' on the basis of their respective performance vectors $\mathbf{g}(a) = [g_1(a), g_2(a), \ldots, g_n(a)]$ and $\mathbf{g}(a') = [g_1(a'), g_2(a'), \ldots, g_n(a')]$ measured on a set of evaluation criteria. A detailed review of these methodologies can be found in Ostanello [1985], Roy [1991], and Vincke [1992].

The problem of establishing a quality profile of a new alternative is very different, once we have defined a set of evaluation criteria and the quality profile of remaining alternatives (inverse decision problem). This is the recurring problem that enterprises have to face when comparing their technical and engineering product performances with competitors on the market.

Which set of technical values, minimizing efforts, can make a product more desirable than competition products? The problem of assigning technical and engineering characteristic values to a product is connected with these issues. To define design targets (quality profile) it is necessary, therefore, to get some tools ready that on the one hand, stimulate the opportunity of occupying a leadership position on the market and, on the other hand, maximize designers' attention toward customer requirements.

From a general point of view the problem can be formulated as follows. Let us define $A = \{a_i \mid i = 1, \ldots, m\}$ as a finite set of potential alternatives, evaluated on a consistent set of criteria $G = \{g_j \mid j = 1, \ldots, n\}$ [Roy, 1991]. Each criterion g_j is considered as a single point application from set A to criterion scale E_j, that is, a completely ordered set (of quantitative or qualitative values) taken as the formal representation of the set of states associated with the j-th criterion:

$$g_j : a \in A \Rightarrow g_j(a) \in E_j$$

Thus, the multiple criteria evaluation of an alternative $a \in A$ can be summarized by the vector $\mathbf{g}(a) = [g_1(a), g_2(a), \ldots, g_n(a)] \in \mathfrak{I} = E_1 \times E_2, \ldots, E_n$.

If $A = \{a_1, a_2, \ldots, a_m\}$ is the set of the compared alternatives, $g_j(a)$ is called the performance of a generic alternative a on j-th criterion. $\forall a', a \in A$ if $g_j(a') \geq g_j(a)$, then a' is at least as good as a if we consider only the point of view reflected by the j-th criterion.

The evaluation of a single alternative has an interesting geometric representation on the criteria space, as is illustrated in Figure 6.1 [Urban and Hauser, 1993]. In this figure, criteria scales and technical characteristic profiles for each alternative are presented. It is important to note that Figure 6.1 does not allow one to establish *at first*

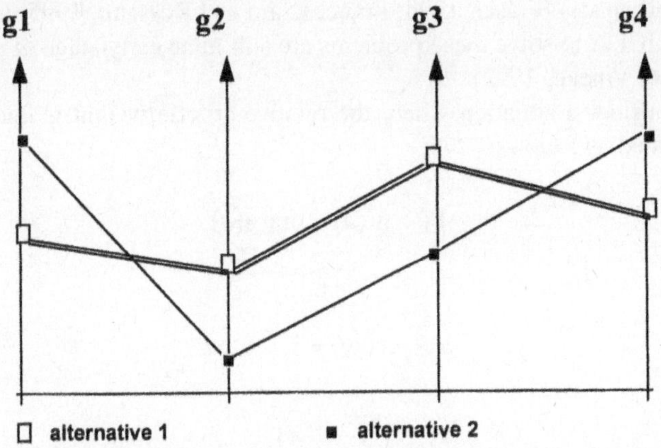

FIGURE 6.1 Representation of alternatives on the criteria space (quality profile).

sight which is the best alternative. In fact, normally, the different criteria's weight is not the same; in other words, the weights are not all equally important. For each criterion, scales are ordered according to an increasing order of preference [Ostanello, 1985].

According to the quality profile concept, the problem of the best alternative selection $\mathbf{g}^*(a) = \left[g_1^*(a), g_2^*(a), \ldots, g_n^*(a)\right]$, in such a way to minimize employed resources, can be formulated as follows:

$$\mathbf{P1}: \mathbf{g}^*(a) = \min_{\Im} \mathbf{g}(a)$$

$$\text{s.t.}$$

$$a >_\psi a_i, \ \forall i = 1, \ldots, m$$

where $\Im = E_1 \times E_2, \ldots, E_n$ represents the search domain defined on the criteria space (design region); $>_\psi$ defines the dominance relation between two alternatives according to a model of preferences ψ; $\mathbf{g}(a)$ is the vector of performances; n is the number of criteria; m is the number of alternatives compared with a; and a_i is the i-th alternative.

Once we have established an aggregation model of preferences able to synthesize the opinion expressed by the decision maker, we can determine a solution of **P1**. The aggregation model is selected on the basis of available data, and of the modelization level with which we are able to describe the decision problem [Ostanello, 1985; Roy, 1991].

In general, **P1** resolution is not easy. **P1** complexity has two main causes: on the one hand, the difficulty in gathering information (definition of alternatives, evaluation criteria, and their importance order, etc.); on the other hand, in some situations, the unavailability of adequate mathematical tools for problem resolution. This is the case, for example, where only ordinal rankings of criteria and alternatives are known. An ordinal scale establishes a priority order among objects; it does not give any indication about their distance in the space of preferences (a *metric* is not defined in this space) [Fraser, 1994; Franceschini and Rossetto, 1995b]. Methods of MCDM or MCDA to solve these problems are still in an early stage of development [Steuer, 1986; Vincke, 1992].

If we consider a situation where the relative criteria weight is known, **P1** can be reformulated as follows:

$$\mathbf{P2}: \ \mathbf{g}^*(a) = \min_{\Im} \mathbf{g}(a)$$

s.t.

$$a >_{\psi} a_i, \ \forall i = 1, \ \ldots, \ m$$

$$\sum_{j=1}^{n} w_j = 1$$

where $\Im$ represents the search domain defined on the criteria space (design region); $>_{\psi}$ defines the dominance relation between two alternatives according to a model of preferences ψ; $\mathbf{g}(a)$ is the vector of performances; w_j is the weight corresponding to the j-th criterion; n is the number of criteria; m is the number of alternatives compared with a; and a_i is the i-th alternative.

Criteria weights to compare alternatives are used in the aggregation model of preferences [Vincke, 1992]. **P1** and **P2** represent an inverse formulation of a typical decision problem solvable with MCDA tools. A new decision problem becomes that of designing a quality profile of an alternative that dominates the other ones, minimizing the employment of available resources.

An approximate solution of **P2** can be determined by means of heuristic algorithms. In the Appendix we propose a scheme of a heuristic algorithm (called *Qbench*) directed toward this goal. An application of Qbench is illustrated by means of an example.

In some particular situations unit costs, associated with each evaluation criteria, are known. In these cases **P1** becomes:

$$\mathbf{P3}: \ \min_{\Im} G = \sum_{j}^{n} c_j x_{aj} w_j$$

s.t.

$$\sum_{j}^{n} c_j x_{aj} w_j \geq_{\psi} \sum_{j}^{n} c_j x_{ij} w_j \quad \forall i = 1, \ \ldots, \ m$$

$$\sum_{j=1}^{n} w_j = 1$$

$$x_{ij} \in E_j \quad \forall i = 1, \ \ldots, \ m$$

where $\Im$ represents the search domain defined on the criteria space (design region); E_j is the j-th criterion scale; c_j is the cost for criterion units; w_j is the weight corresponding to the j-th criterion; n is the number of criteria; m is the number of alternatives compared with a; x_{ij} is the number of criterion units for each alternative ($i = 1, \ldots, m$); and x_{aj} is the number of criterion units for the new alternative.

The **P3** aggregation model is defined by an overall weighted cost function. In such a way the original problem **P1** is taken back to the resolution of a mixed integer linear programming in x_{aj}.

6.3 THE PRODUCT–PENCIL EXAMPLE

Let us again consider a simplified example of a design of a pencil. We make use of QFD to address the design toward real needs of the customers. The designer's attention *is driven* by a set of weights that reflect customer prioritization of the technical and engineering product characteristics (see Chapter 4).

On the basis of QFD results and competitors' product performances, the designer must select the best set of technical and engineering characteristic values with which to carry out the product (design targets or target settings).

Interpreting technical and engineering characteristics of a product as evaluation criteria $G = \{g_j \mid j = 1, \ldots, n\}$, each one characterized by a well-defined degree of importance w_j, and different competitors on the market as alternatives $A = \{a_i \mid i = 1, \ldots, m\}$, we have to establish design targets of a new product–pencil (**P2** problem), minimizing efforts and resources.

By reasoning on alternative quality profiles we can identify an infinite set of solutions to our problem that satisfy the condition $a >_\psi a_i$, $\forall i = 1, \ldots, m$ (of **P2**); one of these is, for example, that which assumes the maximum value on each criterion (envelope), and obviously all that dominate it.

Among these infinite solutions, we search for that which reduces assumed values over each single criterion. Identifying **P2** solutions is not a trivial problem.

Figure 6.2 illustrates the house of quality (HoQ) for the product–pencil. In this context customer needs are ***b*1** "easy to hold," ***b*2** "does not smear," ***b*3** "point lasts," and ***b*4** "does not roll."

With reference to these needs, designers identify the following technical and engineering characteristics: ***y*1** "length of pencil," ***y*2** "time between sharpening," ***y*3** "lead dust generated," and ***y*4** "hexagonality."

Pencil hexagonality is measured by means of an indicator (variable between 0 and 1), able to quantify pencil geometric characteristics against an ideal one (the greater the value of the indicator is, the closer the pencil section approximates an ideal hexagon).

Relationships between customer needs and technical and engineering characteristics are illustrated in the *relationship matrix* ***R***. Symbols drawn in matrix ***R*** identify the type of connection between objects.

Symbols (r_{ij} coefficients) contained in ***R*** are converted into numerical values according to the *independent scoring method* (see Chapter 4). The absolute weight w'_j

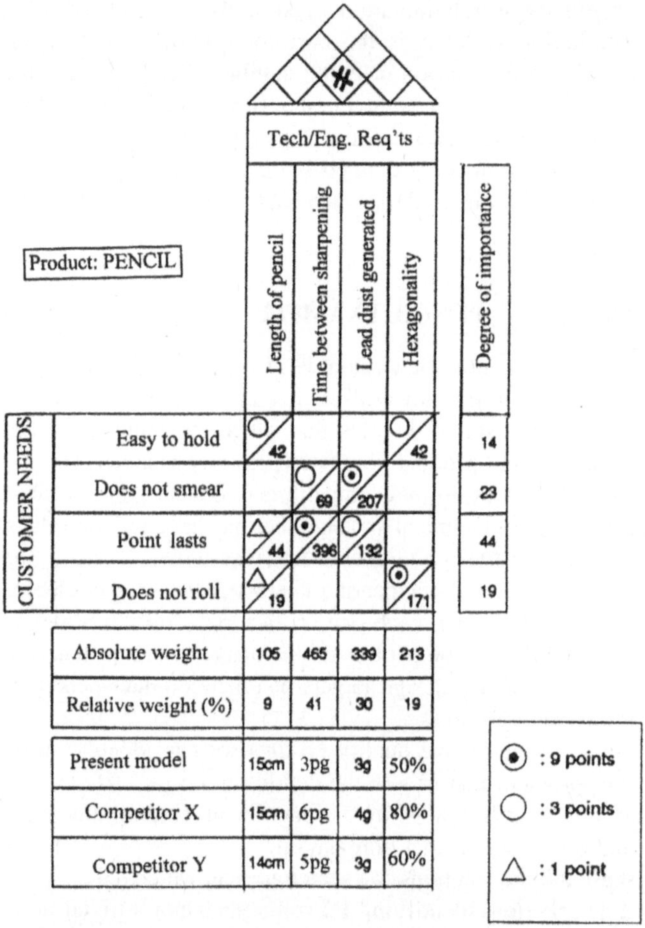

FIGURE 6.2 The house of quality for the pencil example.

of the technical and engineering characteristics are obtained as follows (see Chapter 4, Section 4.6):

$$w'_j = \sum_{i=1}^{k} d_i \cdot r_{i,j}$$

where d_i is the degree of importance of i-th customer requirement, $i = 1, 2, \ldots, k$; k is the number of customer requirements; $r_{i,j}$ is the *quantified* relationship between customer requirement i-th, and engineering design characteristic j-th, $i = 1, 2, \ldots, k$, $j = 1, 2, \ldots, n$; and w'_j is the technical importance rating for engineering design characteristic j-th, $j = 1, 2, \ldots, n$.

The relative normalized weight of the engineering design characteristic w_j can be determined as follows:

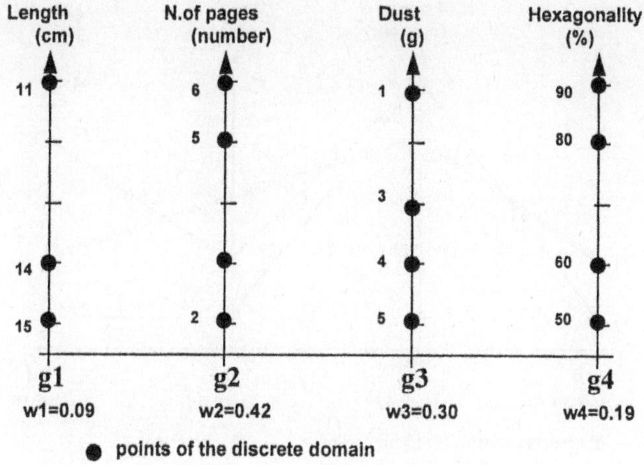

FIGURE 6.3 Search domain of the proposed algorithm.

$$w_j = \frac{w'_j}{\sum_{i=1}^{n} w'_i}$$

$$j = 1, 2, \ldots, n.$$

Conversion of $r_{i,j}$ coefficients has been carried out assigning nine points to the strong relationships, three points to the medium relationships, and one point to the weak relationships [Wasserman, 1993]. For the opportunity of the conversion process of ***R*** coefficients see Chapter 7. Weight numerical values are reproduced in Figure 6.2.

At this point the problem of establishing the technical features of a product arises, considering customer requirements and competitor profiles. A solution to the problem is given by the Qbench algorithm discussed in the Appendix.

For aggregation, the ELECTRE II method has been considered (see selection criteria proposed in Ostanello [1985] and Roy [1991]).

Extreme values on each criterion scale are:

$\mathbf{y1}: [e_1^i, e_1^s] \rightarrow [15, 11]$ cm $\qquad \mathbf{y2}: [e_2^i, e_2^s] \rightarrow [2, 6]$ number of pages

$\mathbf{y3}: [e_3^i, e_3^s] \rightarrow [5, 1]$ g $\qquad \mathbf{y4}: [e_4^i, e_4^s] \rightarrow [50, 90]$ %

Criteria scales are drawn according to an increasing decision maker's preference order. The selection of extreme values on each scale is carried out on the basis of technological and economic constraints.

The competitor X (alternative X) quality profile is represented by the vector [15, 6, 4, 80], whereas the competitor Y (alternative Y) quality profile by the vector [14, 5, 3, 60].

The solution search domain is illustrated in Figure 6.3. As one can observe, the search domain is constituted by a finite set of points. The present solution obtained

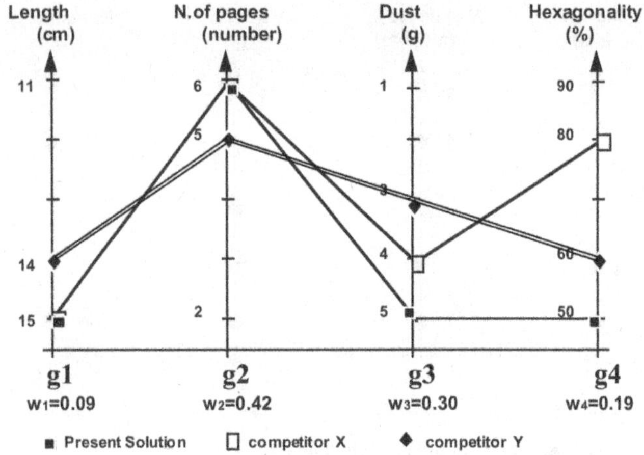

FIGURE 6.4 Competitors' profiles and initial solution proposed by the algorithm.

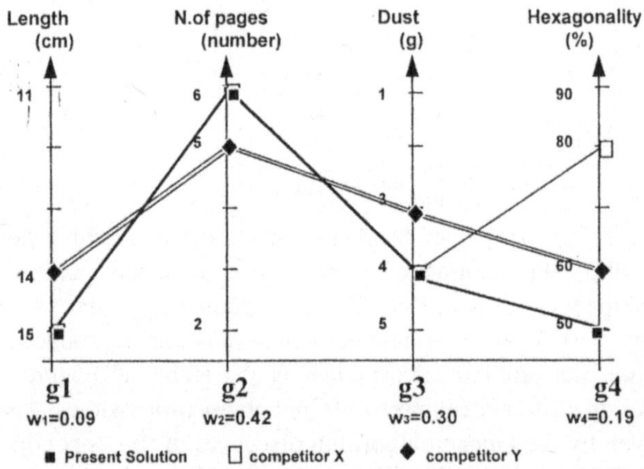

FIGURE 6.5 Competitor profiles and present solution: step1.

applying the Qbench algorithm, and alternative X and Y quality profiles are visualized in Figure 6.4. The present solution produced by the algorithm is then subject to the verification test, which is not met.

The algorithm generates a new solution advancing one step along **y3** criterion, having already achieved the maximum on **y2** criterion. The new situation is illustrated in Figure 6.5. This solution does not meet the verification test, point (b) (see Appendix).

Another step is then carried out on **y3** criterion. The new vector is [15, 6, 3, 50], which now does pass the point (b) test, reaching the final solution (see Figure 6.6).

6.4 RESULTS AND OBSERVATIONS

From the pencil example we can realize the easiness and quickness of application of the Qbench algorithm. One of the most important ideas behind Qbench is the

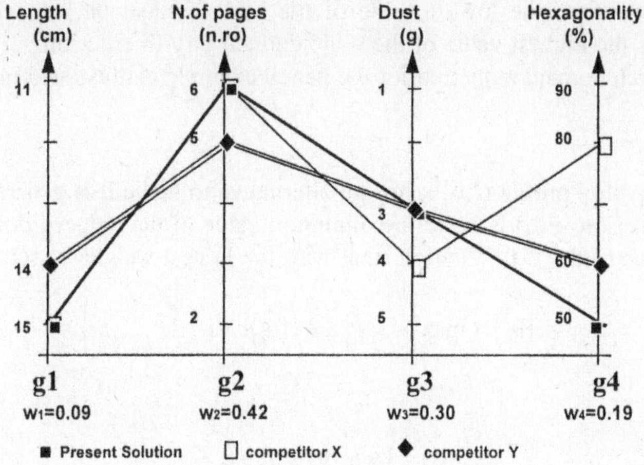

FIGURE 6.6 Competitors' profiles and final solution proposed by the algorithm.

search domain reduction, which has an immediate impact on the time spent to find a solution. On the other hand, the hypothesis of the search domain reduction is completely aligned with a general procedure of defining design targets with reference to competition, moving technical and engineering targets in a limited domain compatible with technological and economic constraints.

It is important to observe that the Qbench algorithm can be utilized independently by the particular model employed to carry out the preference aggregation. In this sense, the verification test discussed at point (b) can be implemented, for example, by the analytical hierarchy process (AHP) method (see Chapter 5) when we have only cardinal criteria or others [Dyer and Sarin, 1979].

Appendix — Qbench Algorithm

Assumptions. Let us suppose that a problem's solution can assume only values coincident with the criteria's extreme scale positions, or values taken from other alternatives on each criterion (search domain reduction):

$$g(a) \in \Omega \subseteq \Im$$

where

$$\Im = E_1 \times E_2, \ \dots, \ E_n$$

$$\Omega = H_1 \times H_2, \ \dots, \ H_n$$

$$H_i \subseteq E_i \ \forall i$$

$$H = \left\{ e_j^i, \ e_j^s, \ g_j(a), \ \forall i = 1, \ \dots, \ m; \quad j = 1, \ \dots, \ n \right\}$$

$$J = \{1, \ \dots, \ n\}, \ I = \{1, \ \dots, \ m\}$$

where e_j^i represents the lowest value of the scale defined on j-th criterion; and e_j^s represents the highest value of the scale defined on j-th criterion.

The search domain reduction for the pencil example is illustrated in Figure 6.3.

INITIALIZATION

The initial quality profile (l = 1) for the alternative to be built is generated. For each criterion, $g_j(a)$ is set at the minimum value of the reduced domain H.

The value of the j^*-th criterion (that with the largest weight) is set to:

$$g_{j^*}(a^1) = \max\left\{e_{j^*}^i,\ e_{j^*}^s,\ g_{j^*}(a_i),\ \forall i = 1,\ \ldots,\ m\right\}$$

$$l = 1$$

$$g_j(a^l) = e_j^i\ \forall j \in J,\ j \neq j^*$$

$$j^* \big| w_{j^*} = \max_{j \in J} w_j$$

$$g_{j^*}(a^l) = \max\left\{e_{j^*}^i,\ e_{j^*}^s,\ g_{j^*}(a_i),\ \forall i = 1,\ \ldots,\ m\right\}$$

$$J^* = J - \{j^*\}$$

go to (b).

Figure 6.4 shows the initial solution for the pencil example. The criterion with the largest weight is j^* = 2 (number of pages; w_2 = 0.42). The global evaluation vector for this solution is $\mathbf{g}(a^1)$ = [15, 6, 5, 50], with $g_2(a^1)$ = 6.

(a) GENERATION OF THE l-th (l > 1) QUALITY PROFILE

Generation of the l-th quality profile of the searched alternative is as follows.

if

$$g_{j^*}(a^{l-1}) = e_{j^*}^s$$

then

$$j^* \big| w_{j^*} = \max_{j \in J^*} w_j$$

$$g_j(a^l) = g_j(a^{l-1})\ \forall j \in J^*,\ i \neq j^*$$

$$g_{j^*}(a^l) = g_{j^*}(a^{l-1}) + \Delta_{j^*}$$

$$J^* = J - \{j^*\}$$

go to (b).

else

$$g_j(a^l) = g_j(a^{l-1}) \ \forall j \in J^*, \ i \neq j^*$$

$$g_{j^*}(a^l) = g_{j^*}(a^{l-1}) + \Delta_{j^*}$$

go to (b).

Figure 6.5 shows the second solution generated by the algorithm for the pencil example. At the beginning, the new criterion with the largest weight value is determined ($j^* = 3$; "dust" with $w_3 = 0.30$); then, just in this criterion, an advancing step Δ_{j^*} is carried out. The new evaluation vector is $\mathbf{g}(a^2) = [15, 6, 4, 50]$.

(b) Verification Test

If the condition $a^l >_\psi a_i, \forall i = 1, \ldots, m$ is verified, we stop the procedure; else we return to (a).

if

$$a^l >_\psi a_i, \ \forall i = 1, \ldots, m$$

then stop: a^l is the searched solution.
else

$$l = l + 1$$

go to (a).

end

Δ_{j^*} represents the increment of one step on the j-th* criterion.

The algorithm stops when the searched solution is the best from the decision maker's model of preference (managed by means of the ELECTRE II method [Ostanello, 1985; Roy, 1991]). The final solution is represented in Figure 6.6 and the corresponding evaluation vector is $\mathbf{g}(a^3) = [15, 6, 3, 50]$. Figure 6.7 shows a conceptual scheme of the algorithm.

With reference to the algorithm we can observe the following:

- The original problem of searching for a solution in a mixed integer-continuous domain has been converted to an analogous problem with a reduced search domain (discrete domain).
- The Qbench algorithm always gives a solution; in the worst case it coincides with the maxima's envelope.
- Step (a) of Qbench can be substituted for by a breadth search strategy, instead of a depth search as suggested in the procedure.

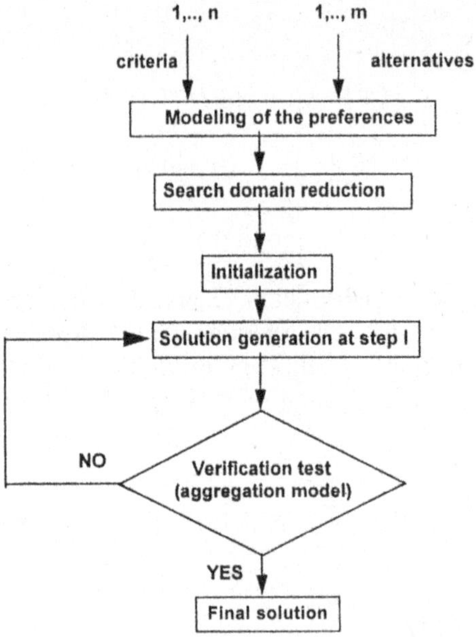

FIGURE 6.7 Conceptual scheme of the proposed algorithm.

- Computational complexity of the algorithm in the worst case is $o(m{\cdot}n)$, with n and m the number of criteria and the number of alternatives, respectively.
- The verification test (b) can be executed only when we have defined the aggregation model of preferences to utilize.

REFERENCES

Akao, Y. (1992), Origins and Growth of QFD, First European Conference on Quality Function Deployment, Milano.

ASI (1987), Quality Function Deployment, Executive Briefing, American Supplier Institute, Dearborn, MI.

Bemowski, K. (1991), The benchmarking bandwagon, *Qual. Prog.*, 24(1), 19–24.

Dyer, J.S. and Sarin, R. (1979), Measurable multiattribute value functions, *Operations Res.*, 27(4), 810–822.

Franceschini, F. (1993), Impostazione di progetti di grande dimensione: il vincolo della Qualità, *Logistica Manage.*, 36, 34–42.

Franceschini, F. and Rossetto, S. (1995a), Quality & innovation: a conceptual model of their interaction, *Total Qual. Manage.*, 6(3), 221–229.

Franceschini, F. and Rossetto, S. (1995b), QFD: the problem of comparing technical/engineering design requirements, *Res. Eng. Design*, 7, 270–278.

Franceschini, F. and Rossetto, S. (1997), Design for quality: selecting product's technical features, *Qual. Eng.*, 9(4), 681–688.

Franceschini, F. and Zappulli, M. (1998), Product's technical quality profile design based on competition analysis and customer requirements: an application to a real case, *Int. J. Qual. Reliability Manage.*, 15(4), 431–442.

Fraser, N.M. (1994), Ordinal preference representations, *Theory Decision,* 36(1), 45–67.

Garvin, D.A. (1987), Competing on the eight dimensions of quality, *Harv. Bus. Rev.,* 65(6), 101–109.

Hauser, J. and Clausing, D. (1988), The house of quality, *Harv. Bus. Rev.,* 66(3), 63–73.

Ostanello, A. (1985), Outranking methods, in *Multiple Criteria Decision Methods and Application*, Fandel, G. and Spronk, J., Eds., Springer-Verlag, Berlin, pp. 41–60.

Roy, B. and Bertier, P. (1972), La méthode ELECTRE II: une application au media-planning, VIIe Conférence Internationale de Recherche Opérationelle, Dublin.

Roy, B. (1990), Decision-aid and decision-making, *Eur. J. Operational Res.*, 45, 324–331.

Roy, B. (1991), The outranking approach and the foundations of ELECTRE methods, *Theory Decision,* 31(1), 49–73.

Steuer, R. (1986), *Multiple Criteria Optimization: Theory, Computation and Application*, John Wiley & Sons, New York.

Urban, G.L. and Hauser, J.R. (1993), *Design and Marketing of New Products*, Prentice Hall International, Englewood Cliffs, NJ.

Vincke, P. (1992), *Multicriteria Decision Aid*, John Wiley & Sons, Chichester.

Wasserman, G.S. (1993), On how to prioritize design requirements during the QFD planning process, *IIE Trans.,* 25(3), 59–65.

Zairi, M. (1992), The art of benchmarking: using customer feedback to establish a performance gap, *Total Qual. Manage.*, 3(2), 177–188.

7 The Prioritization of Technical and Engineering Design Characteristics

7.1 INTRODUCTION

Previous chapters have highlighted the different features of quality during the design of a new product. Quality is characterized by the following attributes:

- *Multidimensionality.* It involves more dimensions in the evaluation of a product, for example, performance, features, reliability, conformance, durability, serviceability, security, aesthetics, and perceived quality [Garvin, 1987].
- *Relativity.* Its value is not compared with an absolute value, but with what the customer has perceived.
- *Dynamicity.* Its value varies over time.
- *Globality.* It involves every internal and external function of the company (the economic impact of an activity on a product design decreases with its distance over time from the commencement of the project).

For these reasons, good structuring of a new project up to its preliminary phases (involving the customer as a primary "actor") is of fundamental importance.

Product designers need to know how to make trade-offs in the selection of design features that result in the highest level of customer satisfaction. Because of the complexity of the decision process, the design team will often rely on *ad hoc* decision procedures to assist in this product development. Such procedures are often completely arbitrary, however, and subject to the whims of the design team instead of the needs of the customer [Hauser and Clausing, 1988; Wasserman, 1993].

In Chapter 4 we have seen how quality function deployment (QFD) can be utilized in the following ways:

- For gathering product information in a structural way by customers and design work teams
- For analyzing customer expectations and the characteristics of competitive products
- For defining the prioritization of technical and engineering design characteristics for a new product

The ranking of product design and engineering characteristics [Behara and Chase, 1993; Wasserman, 1993] is obtained by means of a numerical coding (cardinalization) of the ordinal information contained in the relationship matrix (see Chapter 4, Section 4.6). The rigid coding of the information contained in the QFD tables presents some drawbacks, however, first with the nonadaptability to all situations [Dyer, 1990; Fraser, 1994], and second with the arbitrariness of turning ordinal information into cardinal information.

More formal approaches are then needed to provide an unforced evaluation of the QFD tables. In this regard, an approach based on multiple criteria decision aid (MCDA) methods will be shown to be useful for capturing the real contents of the decision process.

In this chapter we highlight the risks of forcing product design toward nonnatural "trajectories" connected with the cardinalization of the qualitative information appearing in the QFD tables.

7.2 CONVERSION OF RELATIONSHIP MATRIX COEFFICIENTS

QFD tables are filled in by the design work team and the customer product in such a way as to establish what interactions exist between customer requirements and product characteristics. The relationship matrix links customer requirements with technical design characteristics utilizing an ordinal scale with the following levels:

- Strong relationship
- Medium relationship
- Mild or weak relationship

Up to this point QFD plays a typically organizational role in the design of a product. However, if information contained in the relationship matrix is utilized to steer the designer's attention toward the relative importance of the various characteristics of the product (as seen in terms of the customer's requirements), then QFD may become a decisional supporting tool.

The fundamental approach is to consider customers' requirements (which reflects the priority order expressed by the customers), without in any way violating their intentions. The traditional method normally used to generate a meaningful degree of importance to the customer in the prioritization of technical and engineering design requirements makes use of the AHP 1, 3, 9 or 1, 5, 9 scale to denote weak, medium, and strong relationships between pairs of customer and technical design characteristics (see Chapter 5).

With this method the coefficients of the relationship matrix are turned from an ordinal to a cardinal scale. At this point the question arises about the legitimacy of the transformation of coefficients $\boldsymbol{r}_{i,j}$ contained in relationship matrix $\boldsymbol{R}$. Up to which level does the numerical conversion in $\boldsymbol{R}$ conform to the assurance that the customer requirements are actually met?

The process of converting ordinal information into a cardinal scale represents an arbitrary passage (which may sometimes be dangerous). In fact, it produces a double negative effect: on the one hand, it assigns an arbitrary precision to customer assessments, whereas, on the other hand, it imposes an element of absolute truth on the customer evaluations, imparting to them more informative content than has in fact been expressed.

An ordinal scale establishes a priority order among objects; it does not give any indication about their distance in the ranking. The operations of weighting and adding are not meaningful, because it is easy to produce different results by choosing different scales from which to draw the ordinals [Saaty, 1990]. A cardinal scale not only establishes an order of precedence among objects but also gives a score to each of them. Thus, besides giving the ranking order it indicates the distances between ranks in terms of differences between their acquired scores [Fraser, 1994].

In a cardinal scale (unlike an ordinal one) a *metric* for the ranking evaluation of *n* objects is then defined. Turning from one to another is therefore questionable [Harker and Vargas, 1987]; it implies, independently of the particular method utilized, further specifications or information. Incidentally, it is not trivial to note that if we had carried out for our example an ordinal–cardinal conversion using another numerical scale (e.g., 1, 3, 5), we would have obtained a ranking of technical and engineering design characteristics different from the previous one (so there is low robustness to the variation of the values of the cardinal scale elements).

Difficulties in the organization of a product design, in the evaluation of the correct level of design detail, and in the lack of familiarity with the innovative product characteristics are only a few of the reasons that make questionable the rigid conversion (by means of a cardinal scale) of the information contained in the relationship matrix. On the contrary, QFD tables if used as decision supporting tools must be easy to use, and faithful to the *real decision mechanisms* that drive the development of a product. The main objective of such a supporting tool is to construct or to create some ability to help individual actors taking part in the decision process, either to argue or to transform their (or the customers') preferences or to make decisions in conformity with their goals of:

- Extracting what appears to be really meaningful from the available information (in the perspective of what needs to be established)
- Helping the decision makers' behavior by bringing to them arguments that can strengthen or weaken their own convictions

If the conversion of the relationship matrix coefficients from an ordinal scale to a cardinal one produces some hesitation by the designers, what is the remedy for establishing a priority order for the technical and engineering design characteristics of a product? In other words, is it possible to conceive an aggregation model of the customer's requirements (preferences) based on principles different from those of a cardinal (enforced) coding of the information contained in the relationship matrix?

Methods developed in the MCDA environment allow us to face these issues. MCDA efforts are oriented toward concepts, properties, and procedures, which are liable to be used for the previous purposes [Roy, 1990; Vansnick, 1986a, 1986b].

7.3 QUALITY FUNCTION DEPLOYMENT AND MULTIPLE CRITERIA DECISION AID

The conditions that can influence the building of a technical importance ranking for engineering design characteristics are:

- The nature of basic concepts used to rank requirements (concordance, discordance, substitution rate, intensity of preference, etc.)
- The degree of significance of the customer requirements that are to be taken into account
- The strength of the arguments required for validating the judgment that one engineering design characteristic is more important than another

These issues are dealt with well by MCDA methods. To understand what they are and what kind of real problems they refer to, it is necessary to specify what is supposed to be given initially:

- A set $A = \{a_1, a_2, \ldots, a_n\}$ of potential *alternatives* is given. (Such alternatives are not necessarily exclusive, i.e., they can be put into operation jointly.)
- A consistent set of *criteria* g_j is provided. (The preferences of actors involved in the decision process are argued and transformed by reference to points of view adequately reflected by criteria g_j [Roy, 1991].)
- The criterion $g_j(a)$ is called the j-th performance of a generic alternative a. It is not restrictive to suppose that $\forall a' \in A$ and $a \in A$, $g_j(a') \geq g_j(a) \Rightarrow a'$ is at least as good as a if we consider only the point of view reflected by the j-th criterion.
- The comparison of a' and a on the basis of the vector of performances $\mathbf{g}(a) = [g_1(a), g_2(a), \ldots, g_n(a)]$ and $\mathbf{g}(a') = [g_1(a'), g_2(a'), \ldots, g_n(a')]$ is done by means of the comprehensive model of preferences able to take into account hesitations between two of the three following cases:

    - $a'Ia$: a' indifferent to a
    - $a'Pa$: a' strictly preferred to a
    - aPa': a strictly preferred to a'

    The hesitations previously mentioned may come from [Roy, 1991]:

    - The existence in the decision maker's mind of zones of uncertainty, conflicts, or contradictions
    - The imprecise and uncertain determination of the vectors $\mathbf{g}(a)$ and $\mathbf{g}(a')$ by means of which a' and a are compared
    - The fact that the builder of the model ignores, in part, how the decision maker compares a' and a

The element at the basis of MCDA is the outranking concept. An outranking relation is a model to aggregate the n criteria of a family $\mathbf{g}$, built with a fewer hypothesis than those required by a value or utility function [Dyer and Sarin, 1979; Ostanello, 1985; Roy, 1991].

Remember that $A = \{a_i \mid i \in I\}$ is a finite set of potential alternatives, evaluated on a consistent set of criteria, $g = \{g_i \mid i \in J\}$. Each criterion g_j is considered as a single point application from the set A to the criterion scale E_j, that is, a completely ordered set (of quantitative or qualitative values) taken as the formal representation of the set of states associated with the j-th criterion:

$$g_j : a \in A \Rightarrow g_j(a) \in E_j$$

Thus, the multiple criteria evaluation of an alternative $a \in A$ can be summarized by the vector $\mathbf{g}(a) = [g_1(a), g_2(a), \ldots, g_n(a)] \in E = E_1 \times E_2, \ldots, E_n$.

The model of the outranking relation consists of admitting that for any pair of alternatives (a, a') of A, *a outranks a'*: (aSa') when both a concordance test and a nondiscordance test are satisfied.

7.3.1 Concordance Test

The concordance test is a measurement of the degree of concordance of the different criteria with the assertion aSa'. The j-th criterion is in concordance with the assertion aSa' if and only if aS_ja'. The subset of criteria that are in concordance with the assertion aSa' is called the *concordant coalition*. Each criterion intervenes in the definition of the coalition strength by means of its weight. Let w_j, $j \in J$, be the weight (positive number) corresponding to the importance given to the j-th criterion within the family $\mathbf{g}$, and denote by $w = \{w_j \mid j \in J\}$ the set of weights.

For any pair of alternatives a and a', the set of criteria for which a is strictly preferred to a' is denoted as $J^+(a, a') \subseteq J$ with

$$J^+(a, a') = \{j \in J : g_j(a) > g_j(a')\}$$

The set of criteria where a and a' get equal evaluations is denoted as $J^=(a, a') \subseteq J$ with

$$J^=(a, a') = \{j \in J : g_j(a) = g_j(a')\}$$

and the set of criteria for which a' is strictly preferred to a is denoted as $J^-(a, a') \subseteq J$ with

$$J^-(a, a') = \{j \in J : g_j(a) < g_j(a')\}$$

The concordance test consists of verifying that the relative importance of the three sets is compatible with the hypothesis aSa'.

When the set of criteria deals with enough differentiated aspects of the consequences, one may simply define the importance of the previously defined subsets by, respectively:

$$W^{+}(a, a') = \sum_{j \in J^{+}} w_j$$
$$W^{-}(a, a') = \sum_{j \in J^{-}} w_j \qquad (7.1)$$
$$W^{=}(a, a') = \sum_{j \in J^{=}} w_j$$

Then a possible formulation of the concordance test may be:

$$c(a, a') = \frac{W^{+}(a, a') + W^{=}(a, a')}{W} \geq k$$

and

$$\frac{W^{+}(a, a')}{W^{-}(a, a')} \geq 1 \qquad (7.2)$$

where $W = \sum_{j \in J} w_j$ and k is a parameter, $0 \leq k \leq 1$, representing the minimum level of concordance or majority. It is supposed that this is discussed with the decision maker; otherwise it is assumed at the so-called natural levels ($k_s = 3/4$; $k_w = 2/3$). k is defined according to the level of uncertainty with which customer information is known [Ostanello, 1985; Roy, 1991]. $c(a, a')$ represents a concordance index; it characterizes the strength of the arguments able to validate the assertion aSa'.

7.3.2 Nondiscordance Test

The nondiscordance test is a measurement of the degree of nondiscordance of the different criteria with the assertion aSa'. The j-th criterion is in discordance with the assertion aSa' if and only if $a'S_j a$. The subset of criteria in discordance with the assertion aSa' is called the *nonconcordant coalition*. Each criterion intervenes in the definition of the nonconcordant strength by means of its weight.

The nondiscordance test allows the management of situations in which the strength of the opposition of certain criteria can be more or less compatible with the acceptance of the assertion aSa'. For reflecting the capacity of a single criterion to reject the assertion aSa' without any help of other criteria, a veto condition is also introduced. This veto effect works on the principle of all or nothing.

The nondiscordance test is introduced to represent veto situations so as to inspect the relative positions of two compared alternatives on the value scales, for those criteria that are in discordance with the hypothesis aSa', that is, the criteria of $J^-(a, a')$.

Depending on the nature of the scale (continuous or not, quantitative or qualitative) and on the decision maker's ability to identify possible veto situations, cardinal or ordinal forms can be adopted [Ostanello, 1985].

A nondiscordance index may be defined, for example, as:

$$d(a, a') = \begin{cases} 0 & \text{if } J^-(a, a') = \varnothing \\ \left(\frac{1}{f}\right) \max\limits_{j \in J^-}\left[g_j(a') - g_j(a)\right] & \text{if } J^-(a, a') \neq \varnothing \\ 1 & \text{if } \exists j \in J^-(a, a') / \text{the veto effect is activated} \end{cases} \tag{7.3}$$

where $f = \max\limits_{j \in J^-}\left(e_{Mj} - e_{mj}\right)$, with e_{Mj}, e_{mj} upper and lower state of E_j, respectively, $\forall j \in J$; the nondiscordance test may be formulated as:

$$d(a, a') \leq q$$

$$0 \leq q \leq 1 \tag{7.4}$$

where q is a threshold, representing a maximum tolerance level of relative negative deviation, consistent with the hypothesis aSa' [Roy, 1991].

7.3.3 Pencil Example

Now we can analyze how MCDA methods can be applied to prioritize design and engineering characteristics in the QFD approach.

With reference to the example explained in Figure 4.10, the ranking order among the symbols that define the different relationships between the customer requirements and the design characteristics may be expressed as follow:

$$\odot > \bigcirc > \triangle$$

The symbol > must be interpreted as the "more important than" operator, whereas the symbol ~ is the "as important as" operator. With this notation the information contained in the QFD table can be rewritten as (for the $r_{i,j} = 0$ coefficients contained in the relationship matrix $\boldsymbol{R}$, where any relation does not appear, a dummy relationship ♦ has been considered with $\triangle >$ ♦):

$$\begin{array}{lll} \mathbf{g}_1\text{: (Easy to hold):} & a_4 > a_1 > a_2 \sim a_3 \sim a_5 & w_1 = 0.17 \\ \mathbf{g}_2\text{: (Does not smear):} & a_5 \sim a_3 > a_2 > a_1 \sim a_4 & w_2 = 0.25 \\ \mathbf{g}_3\text{: (Points lasts):} & a_3 \sim a_5 > a_2 > a_1 > a_4 & w_3 = 0.41 \\ \mathbf{g}_4\text{: (Does not roll):} & a_4 > a_1 > a_2 \sim a_3 \sim a_5 & w_4 = 0.17 \end{array} \tag{7.5}$$

TABLE 7.1
First Phase of the ELECTRE II Method for the Prioritization of the Technical Characteristics of the Pencil Example

(a, a′)	$J^+(a, a')$	$J^+(a, a')$	$J^+(a, a')$	$\frac{W^+ + W^-}{W}$	$\frac{W^+}{W^-} \geq 1$	ASa' (k = 0.66)
(a_1, a_2)	{1, 4}	—	{2, 3}	0.34	No	No
(a_1, a_3)	{1, 4}	—	{2, 3}	0.34	No	No
(a_1, a_4)	{3}	{2}	{1, 4}	0.66	Yes	Yes
(a_1, a_5)	{1, 4}	—	{2, 3}	0.34	No	No
(a_2, a_1)	{2, 3}	—	{1, 4}	0.66	Yes	Yes
(a_2, a_3)	—	{1, 4}	{2, 3}	0.34	No	No
(a_2, a_4)	{2, 3}	—	{1, 4}	0.66	Yes	Yes
(a_2, a_5)	—	{1, 4}	{2, 3}	0.34	No	No
(a_3, a_1)	{2, 3}	—	{1, 4}	0.66	Yes	Yes
(a_3, a_2)	{2, 3}	{1, 4}	—	1	Yes	Yes
(a_3, a_4)	{2, 3}	—	{1, 4}	0.66	Yes	Yes
(a_3, a_5)	—	{1, 2, 3, 4}	—	1	Yes	Yes
(a_4, a_1)	{1, 4}	{2}	{3}	0.59	No	No
(a_4, a_2)	{1, 4}	—	{2, 3}	0.34	No	No
(a_4, a_3)	{1, 4}	—	{2, 3}	0.34	No	No
(a_4, a_5)	{1, 4}	—	{2, 3}	0.34	No	No
(a_5, a_1)	{2, 3}	—	{1, 4}	0.66	Yes	Yes
(a_5, a_2)	{2, 3}	{1, 4}	—	1	Yes	Yes
(a_5, a_3)	—	{1, 2, 3, 4}	—	1	Yes	Yes
(a_5, a_4)	{2, 3}	—	{1, 4}	0.66	Yes	Yes

g_1, g_2, g_3, g_4 are evaluation criteria (in this case they correspond to the customer requirements expressed by the customer–decision maker) with which to select the alternatives corresponding to the technical and engineering product design characteristics: a_1 (length of pencil), a_2 (time between sharpening), a_3 (lead dust generated), a_4 (hexagonality), a_5 (minimal erasure residue), and with w_i, $i = 1, 2, \ldots, n$ the weight of the single criterion.

The problem of establishing a ranking for technical characteristics has then led to an MCDA process. Among various methods developed in the MCDA environment, it is possible to select one that is more adequate to face the problem in Equation (7.5). Our specific example has been solved by adopting the ELECTRE II method [Roy, 1991]. This method is structured in two operative phases:

1. The first one (Table 7.1) is the modeling of the outranking relation aSa'. We say that a outranks a' if and only if $c(a, a') \geq k$ and $\nexists\, j \in J$: $[g_j(a), g_j(a')] \in D_j$, where D_j is the nondiscordance conditions set.
2. The second is the building of a ranking of the alternatives based on the progressive relaxation of the outranking relation. This activity is preceded by the construction of an outranking graph $Gr = (A, S)$, in which nodes represent the alternatives and links the outranking relations (Figures 7.1 and 7.2).

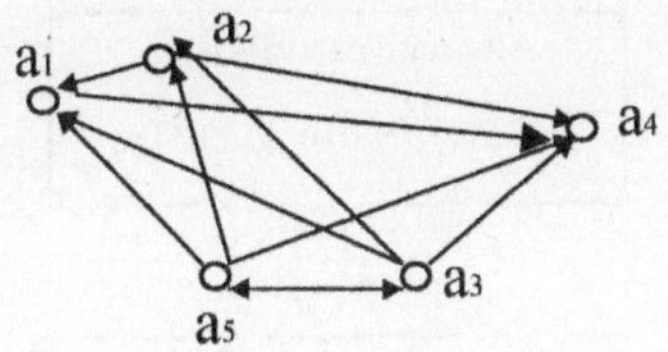

FIGURE 7.1 Outranking graph for the prioritization of the technical characteristics of the pencil example.

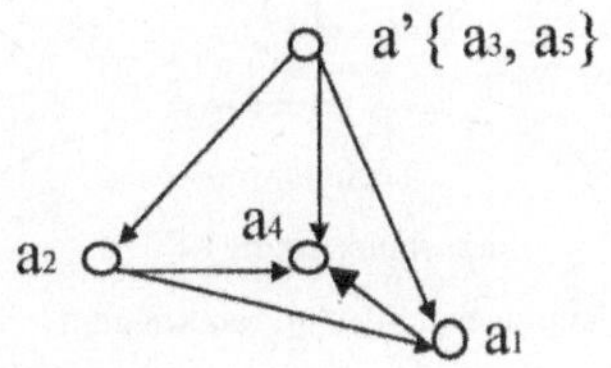

FIGURE 7.2 Reduced-outranking graph for the prioritization of the technical characteristics of the pencil example (see also Figure 7.1).

The ranking obtained is the following:

$$\{a_5, a_3\}, a_2, a_1, a_4$$

Except for the first two alternatives, the ranking obtained for the technical and engineering characteristics does not agree with the one obtained using the traditional rigid conversion method, which on the contrary gives the following:

$$\{a_5, a_3\}, a_4, a_2, a_1$$

These differences can have a considerable effect on the global design of a product (already this trivial example reveals some differences in results between the two methods). In the traditional one the a_4 (hexagonality) characteristic is preferred to the a_2 (time between sharpening) characteristic just in contraposition with what happens on the market where the amount of hexagonal pencils is about the same as the circular ones. On the contrary, the duration (time between sharpening) aspect normally plays a key role in the pencil selection by a customer. It is important to note, therefore, that in the presence of comparable conditions for the main technical and engineering characteristics, the choice to select one or another product (by a customer) is done on the basis of lower importance design characteristics (this is the more common case).

A recapitulatory scheme of the proposed procedure is illustrated in Figure 7.3.

7.3.4 Final Considerations

The comparison of the two analyzed methods (the traditional one — Scoring Method (SM) — and the proposed one — MCDA) allows the following final considerations:

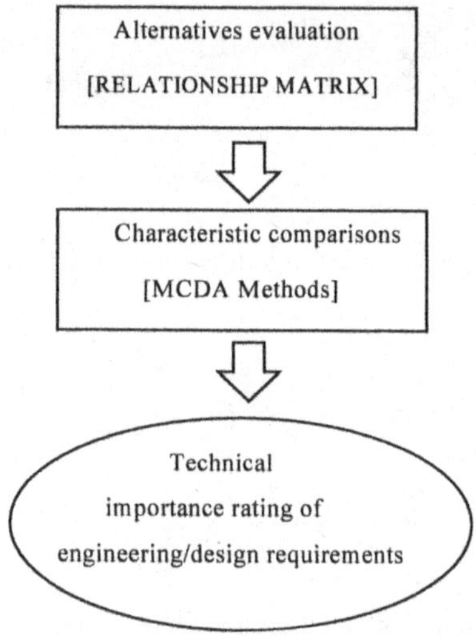

FIGURE 7.3 Recapitulatory scheme of the proposed procedure.

1. The SM method requires a rigid coding (1, 3, 9 scale) of symbols contained in the relationship matrix, which is not adequate for every problem [Harker and Vargas, 1987].
2. The MCDA method does not force out customer evaluations but, on the contrary, takes into consideration the decision mechanisms and the needs of the customer.
3. The MCDA method is able to manage very complex decisional situations, like, for example, veto situations or where it is not possible to express a relative weight among the different criteria of analysis [Roy, 1991]. (The decision maker's preferences very seldom seem well-stated: in and among areas of firm convictions lie hazy zones of uncertainty, half-held belief, or indeed conflicts and contradictions. We have to admit, therefore, that more flexible decisional supports for the QFD tools contribute to answering questions, solving conflicts, transforming contradictions, and destabilizing certain convictions.
4. The MCDA method, when possible, must be utilized in an interactive manner, establishing the concordance and nondiscordance thresholds with customers to meet their requirements. (Data such as numerical values of performances in many cases are imprecise or arbitrary in definition.)
5. The MCDA method is able to manage situations where several people take part in the decision process and where we tend to confuse the one who ratifies the decision with what is called the decision maker.

The proposed method, although it did not give a definitive reply on the effectiveness of QFD in the management of large size projects, seems adequate, however, for the purpose of utilizing approximate customer information without applying rigid and arbitrary conversion of scales. Any complication introduced by MCDA methods is only apparent; in fact, some software packages suitable for their effective employment in a QFD environment are already available on the market [Vincke, 1992].

A method able to partially avoid the problem of forcing out customer evaluations will be illustrated in Chapter 8.

REFERENCES

Akao, Y. (1992), Origins and Growth of QFD, First European Conference on Quality Function Deployment, Milano.

Behara, R.S. and Chase, R.B. (1993), Service Quality Deployment: Quality Service by Design, in *Perspective in Operations Management*, Sarin, R., Ed., Kluwer, Dordrecht, pp. 87–99.

Dyer, J.S. (1990), Remarks on the analytic hierarchy process, *Manage. Sci.*, 36(3), 249–258.

Dyer, J.S. and Sarin, R. (1979), Measurable multiattribute value functions, *Operations Res.*, 27(4), 810–822.

Franceschini, F. (1993), Impostazione di progetti di grande dimensione: il vincolo della Qualità, *Logistica Manage.*, 36, 34–42.

Franceschini, F. and Rossetto, S. (1995), QFD: the problem of comparing technical-engineering design requirements, *Res. Eng. Design*, 7, 270–278.

Franceschini, F. and Rossetto, S. (1997), Design for quality: selecting product's technical features, *Qual. Eng.*, 9(4), 681–688.

Fraser, N.M. (1994), Ordinal preference representations, *Theory Decision,* 36(1), 45–67.

Garvin, D.A. (1987), Competing on the eight dimensions of quality, *Harv. Bus. Rev.*, 65(6), 101–109.

Griffin, A. and Hauser, J. (1992), Patterns of communication among marketing, engineering and manufacturing — a comparison between two new product teams, *Manage. Sci.*, 38(3), 360–373.

Harker, P. and Vargas, L.G. (1987), The theory of ratio scale estimation: Saaty's analytic hierarchy process, *Manage. Sci.*, 33(11), 1383–1403.

Hauser, J. and Clausing, D. (1988), The house of quality, *Harv. Bus. Rev.*, 66(3), 63–73.

Ostanello, A. (1985), Outranking methods, in *Multiple Criteria Decision Methods and Application*, Fandel, G. and Spronk, J., Eds., Springer-Verlag, Berlin, pp. 41–60.

Roy, B. (1990), Decision-Aid and Decision-Making, *Eur. J. Operational Res.*, 45, 324–331.

Roy, B. (1991), The outranking approach and the foundations of ELECTRE methods, *Theory Decision,* 31(1), 49–73.

Saaty, T.L. (1986), Axiomatic foundation of the analytic hierarchy process, *Manage. Sci.*, 32(7), 841–855.

Saaty, T.L. (1990a), Decision making, scaling, and number crunching, in *Decision Making for Leaders: The Analytic Hierarchy Process for Decisions in a Complex World*, (rev. ed.), RWS Publications, Pittsburgh, pp. 269–274.

Saaty, T.L. (1990b), *Multicriteria Decision Making: The Analytic Hierarchy Process*, 2nd ed., RWS Publications, Pittsburgh.

Thurston, D.L. and Locascio, A. (1993), Multiattribute design optimization and concurrent engineering, in *Concurrent Engineering: Contemporary Issues and Modern Design Tools*, Parsaei, H.R. and Sullivan, W.G., Eds., Chapman & Hall, Cambridge, pp. 207–230.

Sullivan, L., (1986), Quality function deployment, *Qual. Prog.*, 19(6), 39–50.

Vansnick, J.C. (1986a), On the problem of weights in multiple criteria decision making (the noncompensatory approach), *Eur. J. Operational Res.*, 24, 288–294.

Vansnick, J.C. (1986b), De borda et condorcet à l'agrégation multicritere, *Ricerca Operativa,* 40, 7–44.

Vincke, P. (1992), *Multicriteria Decision Aid*, John Wiley & Sons, Chichester.

Wasserman, G.S. (1993), On how to prioritize design requirements during the QFD planning process, *IIE Trans.,* 25(3), 59–65.

8 Interactive Design Characteristics Ranking Algorithm for the Prioritization of Product Technical Design Characteristics

8.1 INTRODUCTION

Many firms realize that getting high-quality products to customers in a timely manner is crucial for their survival in a competitive marketplace. Product development process is a complex managerial process that involves multifunctional groups with different points of view as we have seen in the previous chapters. Quality function deployment (QFD) is an innovative tool that stresses cross-functional integration and provides a means of translating product requirements into design specifications [Sullivan, 1986; Hauser and Clausing, 1988; Akao, 1990; Pahl and Beitz, 1996; Ertas and Jones, 1993; Cohen, 1995; Franceschini and Rossetto, 1998; Franceschini, 1998].

Despite its apparent easiness, if information contained in the house of quality (HoQ) is not sufficiently accurate, QFD can become a misleading tool. Its correct and effective use needs careful design analysis and accurate data collection (see Chapter 4, Section 4.2).

After customer identification, the first step of the QFD process is the setting up of procedures for gathering information by customers [Griffin and Hauser, 1992]. The second step concerns data management and elaboration. Typical examples of these activities are the definition of customer requirements and the evaluation of their relative degree of importance. Methods for determining the importance ratings of technical characteristics are dependent on the representation of the symbols contained in the relationship matrix. If symbols are converted in a 1, 3, 9 numerical scale, we may use the simple scoring method [Akao, 1990; Wasserman, 1993, see also Chapter 7]. Such procedures can become arbitrary in those situations in which the customers are not able to give significant evaluation of their requirements and their preference system is not explicitly known. The results of this forcing can lead to a distortion of the design process [Franceschini and Rossetto, 1995; Franceschini and Rupil, 1999, see also Chapter 7]. In fact, customers are forced to give an unnatural evaluation on a conventional scale unnatural to themselves [Larichev et al., 1993, 1995]. At the same time, it is dangerous to carry out an *ex-post* conversion of customer ordinal judgments into numerical scores, because of the introduction of an exogenous and extraneous metric to the formulated judgments [Fraser, 1994; Franceschini and Rossetto, 1995].

The extreme consequences of the use of inadequate conversions can lead to a setting up of a design of a product for an ideal customer that is different from the real one. The soft issue is that we do not know the distance between the two designs.

With specific reference to QFD, the introduction of weights [Vansnick, 1986] to assign a relative degree of importance to customers' requirements can lead to a prioritization order of technical characteristics, which does not reflect their own real intentions. (See methods based on the analytic hierarchy process [AHP] [Akao, 1990; Dyer, 1990, Saaty, 1990]; and also Chapter 5.)

With the aim to better support and facilitate the engineering design process, in this chapter we present an interactive algorithm, which tries to soften customers' approach to QFD. In more detail, it determines a ranking order of design characteristics without the artificial conversion of symbols contained in the relationship matrix, and without explicitly knowing the relative degree of importance of customer requirements.

8.2 RANKING OF TECHNICAL DESIGN CHARACTERISTICS

The traditional QFD approach provides two steps for the ranking of technical design characteristics. The first one concerns the artificial conversion of the relationships between customer requirements and technical design characteristics into numerical equivalent values (see Chapter 4).

The second step provides the determination of relative weights w'_j and normalized relative weights w_j of technical design characteristics (see Chapter 6):

$$w'_j = \sum_{i=1}^{k} d_i \cdot r_{i,j}$$

$$w_j = \frac{w'_j}{\sum_{j=1}^{n} w'_j}$$

$$j = 1, 2, \ldots, n \tag{8.1}$$

Weights so determined represent the *importance* that the customer indirectly ascribes to each technical design characteristic. They can be interpreted as the degree of attention that a designer must reserve for each single technical characteristic during the product development process [Franceschini and Rossetto, 1997; Pahl and Beitz, 1996].

The determination of weights needs knowledge of the degree of importance of each customer's requirements (d_i), and the conversion of symbols contained in the relationship matrix $\boldsymbol{R}$ into equivalent numerical scores ($r_{i,j}$). These are two delicate issues that were explained earlier.

In this section, we are going to present an alternative approach able to manage those situations in which customers are not able to give a cardinal score to the importance of their requirements, without operating an artificial conversion of symbols. The method requires an additional interaction with the customer to dissolve some possible *doubtful* situations, which can arise during the prioritization activity.

The procedure is based on ELECTRE II method [Vincke, 1992; Roy, 1996].

8.3 INTERACTIVE DESIGN CHARACTERISTICS RANKING ALGORITHM

8.3.1 General Assumptions

By adopting the formalism introduced in the previous chapters, let us define $A = \{a_i / i = 1, \ldots, m\}$ as a finite set of potential alternatives (technical characteristics), evaluated on a consistent set of criteria $G = \{g_i / i = 1, \ldots, m\}$ (customer requirements).

Symbols in **R** are not converted in numerical scores. Each coefficient $r_{i,j} \in \mathbf{R}$ is considered as the ordinal evaluation of the j-th alternative by the i-th criterion.

Let us additionally admit that $\forall a', a \in A$, $aS_A a'$ if and only if *concordance* and *nondiscordance* tests are satisfied. On the contrary, $a\,\$_A a'$.

8.3.2 Concordance Test

With reference to the concordance test description (see Chapter 7, Section 7.3.1), we define the set of criteria index as $I = \{1, \ldots, m\}$.

For any pair of alternatives a and a', $I^+\{a, a'\} = \{i \in I\text{: } g_i(a) > g_i(a')\}$ represents the set of criteria for which a is strictly preferred to a', $I^=\{a, a'\} = \{i \in I\text{: } g_i(a) = g_i(a')\}$ is the set of criteria where a and a' get equal evaluations, and $I^-\{a, a'\} = \{i \in I\text{: } g_i(a) < g_i(a')\}$ the set of criteria for which a' is strictly preferred to a.

The concordance test verifies that the relative importance of the three macrosets is compatible with the hypothesis aSa'.

8.3.3 Nondiscordance Test

As in Chapter 7, Section 7.3.2, we introduce a nondiscordance test to take into account eventual *veto* situations. It represents a measurement of the degree of nondiscordance of the different criteria with the assertion aSa'. The i-th criterion is in discordance with the assertion aSa' if and only if $a\,\$_j a'$.

8.3.4 Interactive Procedure

For each pair of alternatives (a, a'), the decision maker, or customer, can express a judgment about the condition $J^=(a, a') \subseteq J$, which produces the relationship aSa', or $a\$a'$. The outranking graph $Gr = (A, S)$ is based on the decision maker's replies. Graph nodes represent alternatives and oriented arcs identify outranking relations.

8.3.5 Ranking Procedure

To generate a ranking of alternatives in the outranking graph a selection procedure is defined. For each iteration $k \geq 1$ a subclass C_k of the final preorder is selected. A_k is the set of alternatives at the k-th iteration.

If

$$k = 1$$

then $A_k = A$ (initialization step)

(a) Generation of the equivalence class at step $k \geq 1$

If

$$|A_k| = 1$$

then $A_k = C_k$ (last class from the top); STOP.

If

$$|A_k| > 1$$

then go to step (b).

(b) Selection of the subset C_k from A_k

Start up the *subprocedure to verify the presence of circuits* in the outranking graph and subsequent graph reduction [Ostanello, 1985; Roy, 1996]:

$$C_k = \left\{ a \in A_k : \nexists a' \in A_k : a' S_{A_k} a \right\}$$

$$A_{k+1} = A_k - C_k$$

If

$$A_{k+1} = \varnothing$$

then STOP.

else go to step (a).

The equivalence class C_k contains the set of elements that outranks the class C_{k-1} and is outranked by the class C_{k+1}.

To have $C_k \neq \varnothing$, S_A must not produce circuits. If some circuits are present, we proceed to a graph $Gr = (A, S)$ contraction. We substitute the circuits with an equivalence class in the graph. Circuits on $Gr = (A, S)$ are identified by means of suitable algorithms from graph theory [Ostanello, 1985; Vincke, 1992].

Figure 8.1 shows a scheme of the interactive design characteristics ranking (IDCR) algorithm. A comparison with the traditional method and a numerical example is provided in the next section.

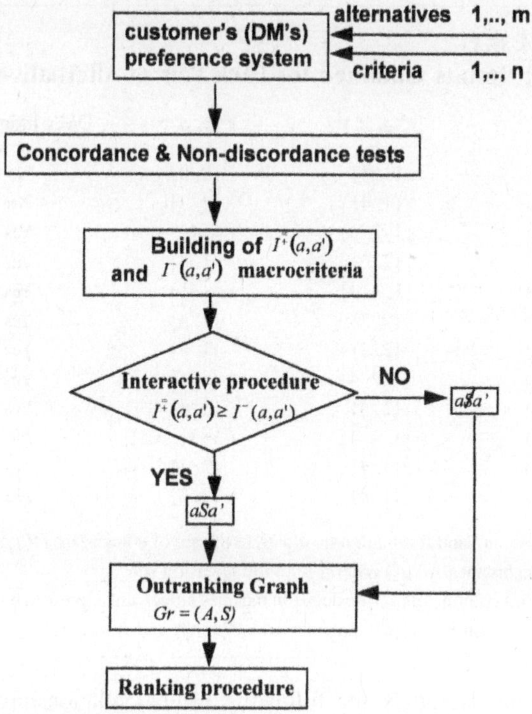

FIGURE 8.1 Conceptual scheme of the IDCR algorithm.

8.4 EXAMPLE OF APPLICATION

Let us consider again the case of a design of a pencil. Figure 6.2 shows the HoQ and the relative relationship matrix. We desire to determine the design characteristics prioritization from the customer point of view. By applying the traditional QFD approach we obtain:

$$Q_2, Q_3, Q_4, Q_1$$

Now we consider the IDCR algorithm.

By interpreting customer requirements as evaluation criteria and product technical design characteristics as alternatives, information contained in the HoQ of Figure 6.2 can be rewritten as:

$$\begin{aligned}
&\mathbf{r}_1\text{: (easy to hold):} && a_1 \sim a_4 > a_2 \sim a_3 \\
&\mathbf{r}_2\text{: (does not smear):} && a_3 > a_2 > a_1 \sim a_4 \\
&\mathbf{r}_3\text{: (point lasts):} && a_2 > a_3 > a_1 > a_4 \\
&\mathbf{r}_4\text{: (does not roll):} && a_4 > a_1 > a_2 \sim a_3
\end{aligned} \tag{8.2}$$

TABLE 8.1
IDCR Results Obtained for Each Pair of Alternatives

(a, a′)	$I^{\mp}(a, a')$[a]	$I^{-}(a, a')$[a]	DM's Judgment[b]
(a_1, a_2)	{1, 4}	{2, 3}	No
(a_1, a_3)	{1, 4}	{2, 3}	No
(a_1, a_4)	{1, 2, 3}	{4}	Yes
(a_2, a_1)	{2, 3}	{1, 4}	Yes
(a_2, a_3)	{1, 3, 4}	{2}	Yes
(a_2, a_4)	{2, 3}	{1, 4}	Yes
(a_3, a_1)	{2, 3}	{1, 4}	Yes
(a_3, a_2)	{1, 2, 4}	{3}	Yes
(a_3, a_4)	{2, 3}	{1, 4}	Yes
(a_4, a_1)	{1, 2, 4}	{3}	No
(a_4, a_2)	{1, 4}	{2, 3}	No
(a_4, a_3)	{1, 4}	{2, 3}	No

[a] The second and third columns illustrate the set of concordant $I^{\mp}(a, a')$ and nonconcordant $I^{-}(a, a')$ criteria with the assertion aSa'.

[b] The last column reports the decision maker's judgement expressed by means of comparison of the two macrocriteria $I^{+}(a, a')$ and $I^{-}(a, a')$.

Matrix $\boldsymbol{R}$ symbols satisfy the following ordinal relationships: ⊙ > ○ > △. Symbols > and ~ must be interpreted as the "more important than" and "as important as" operators, respectively. For the $r_{ij} = 0$ coefficients contained in the $\boldsymbol{R}$ matrix we consider again a dummy relationship ♦ with the condition ♦ > △.

With this new formulation, the original problem is transformed in the determination of the best alternatives ranking, subject to conditions expressed by Equation (8.2).

Table 8.1 contains some intermediate results of the application of the IDCR algorithm. For each pair of alternatives, the second and third columns report the set of concordant $I^{\mp}(a, a')$, and nonconcordant $I^{-}(a, a')$ macrocriteria with the assertion aSa', respectively.

At this point the design team activates the *interactive procedure* with the decision makers (customers). On the basis of both their preference system and the comparison of the two macrocriteria $I^{\mp}(a, a')$ and $I^{-}(a, a')$, the decision makers establish outranking relationships. The obtained results are reported in the last column of Table 8.1. The example does not consider veto situations.

The decision maker's judgments reflect the *implicit* degree of importance of customer requirements. The last column of Table 8.1 allows building the outranking graph $Gr = (A, S)$, as illustrated in Figure 8.2. The graph connects alternatives that satisfy the relation aSa'. If aSa', then $Gr = (A, S)$ contains an arc that links a with a' with the arrow directed to a'. So, for example, for the pair (a_1, a_4) an arc links the node a_1 with the node a_4.

By analyzing $Gr = (A, S)$, we can detect the presence of a circuit between a_2 and a_3. This circuit is due to the mutual outranks of the two alternatives. It becomes

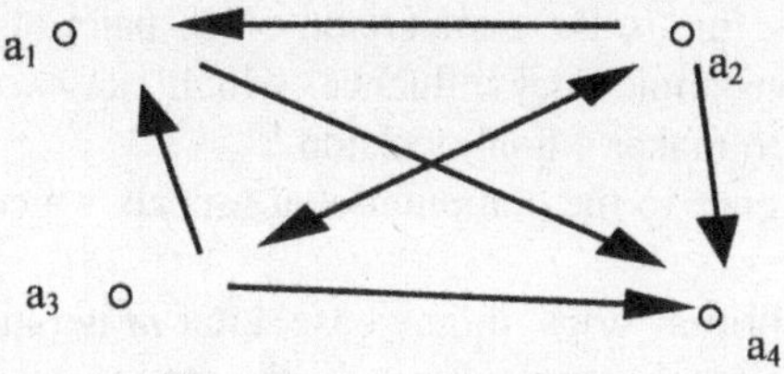

FIGURE 8.2 Outranking graph for the technical design characteristics of a pencil.

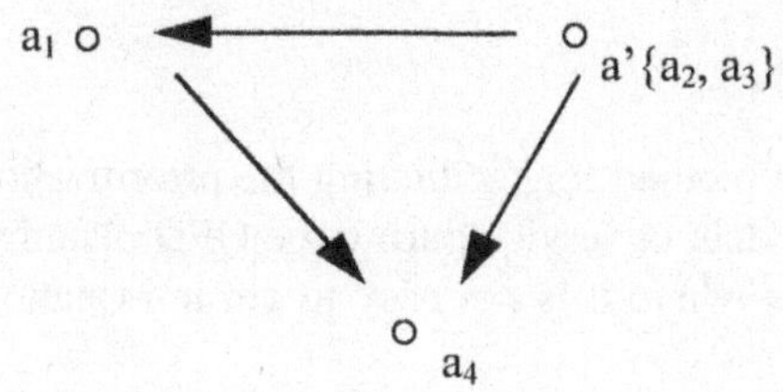

FIGURE 8.3 Outranking graph contraction for the presence of a cycle in the Figure 8.2.

necessary to graph a contraction (Figure 8.3). By applying the IDCR *procedure* the following ranking is obtained:

$$\{a_2, a_3\}, a_1, a_4$$

Alternatives $\{a_2, a_3\}$ belong to the same equivalence class. In this *specific* case, comparing the results with those obtained with the traditional procedure, we notice a good agreement. The only difference is the inversion of the relative order of the two less important characteristics a_4, a_1.

We underline that by using the IDCR procedure we rank the order of design characteristics, without the artificial conversion of symbols contained in the relationship matrix, and without the use of explicit information concerning the relative degree of importance of customer requirements.

8.5 DISCUSSION AND OBSERVATIONS

It must be emphasized that the IDCR algorithm gives a ranking order of design characteristics using only its ability to manage ordinal information. It also avoids the risk of steering the design in an arbitrary way, depending on the conversion scale used to transform the $\boldsymbol{R}$ matrix symbols.

IDCR data are determined by asking customers to express their judgments without forcing them to reason with conventional unfamiliar scales.

A second issue that must be highlighted is that the IDCR algorithm can be easily automated. It can be insertable in generic commercial SW packages, or integrable with other QFD packages [Buede, 1992].

As regards the traditional approach, IDCR bases its operation on a procedure that is not too stiff and restrictive. For example, it allows managing veto situations.

Its apparent heaviness, due to the comparison of all pairs of alternatives, finds its justification in the nonsymmetrical influence, which can exercise the *indifference relation* on the decision maker's final decision.

Finally, with reference to the computational aspects we can observe that:

- The IDCR algorithm stops, in any case, after m iterations.
- IDCR computational complexity is, in the worst case, $o(n^5 + n^2m)$, with m and n the number of criteria and the number of alternatives, respectively.

8.6 CONCLUSIONS

The chapter presents a method for facilitating the prioritization of technical design characteristics of a product or service during the QFD planning process. It is applicable in those contexts where it is not easy to get information or knowledge from the customer.

Based on the interaction with customers (decision makers), the algorithm faces all situations in which the customers are not able to give a score to their requirements on conventional scales. Besides, the algorithm avoids an inappropriate conversion of qualitative information contained in the relationship matrix.

Although the IDCR method determines a spontaneous relationship with customers, it can present some applicability limits when they are not easily achievable (e.g., customers of wide consumption goods).

REFERENCES

Akao, Y. (1990), *QFD: Integrating Customer Requirements into Product Design*, Productivity Press, Cambridge, MA.

Buede, D.M. (1992), Software review: overview of the MCDA software market, *J. Multi-Criteria Decision Anal.*, 1(1), 59–61.

Cohen, L. (1995), *Quality Function Deployment: How to Make QFD Work for You*, Addison-Wesley, Reading MA.

Dyer, J.S. (1990), Remarks on the analytic hierarchy process, *Manage. Sci.*, 36(3), 249–258.

Ertas, A. and Jones, J.C. (1993), *The Engineering Design Process*, John Wiley & Sons, New York.

Franceschini, F. (1998), Quality Function Deployment: uno strumento concettuale per coniugare qualità e innovazione, Ed. Il Sole 24 ORE Libri, Milano.

Franceschini, F. and Rossetto, S. (1995), QFD: the problem of comparing technical/engineering design requirements, *Res. Eng. Design*, 7, 270–278.

Franceschini, F. and Rossetto, S. (1997), Design for quality: selecting product's technical features, *Qual. Eng.*, 9(4), 681–688.

Franceschini, F. and Rossetto, S. (1998), QFD: how to improve its use, *Total Qual. Manage.*, 9(6), 491–500.

Franceschini, F. and Rupil, A. (1999), Rating scales and prioritization in QFD, *Int. J. Qual. Reliability Manage.*, 16(1), 85–97.

Fraser, N.M. (1994), Ordinal preference representations, *Theory Decision*, 36(1), 45–67.

Griffin, A. and Hauser, J. (1992), Patterns of communication among marketing, engineering and manufacturing — a comparison between two new product teams, *Manage. Sci.*, 38(3), 360–373.

Hauser, J. and Clausing, D. (1988), The house of quality, *Harv. Bus. Rev.*, 66(3), 63–73.

Larichev, O.I., Moshkovich, H.M., Mechitov, A.J., and Olson, D.L. (1993), Experiments comparing qualitative approaches to rank ordering of multiattribute alternatives, *J. Multi-Criteria Decision Anal.*, 2(1), 5–26.

Larichev, O.I., Olson, D.L., Moshkovich, H.M., and Mechitov, A.J. (1995), Numerical vs. cardinal measurements in multiattribute decision making: how exact is enough, *Org. Behav. Hum. Decision Processes*, 64(1), 9–21.

Ostanello, A. (1985), Outranking methods, in *Multiple Criteria Decision Methods and Application*, Fandel, G. and Spronk, J., Eds., Springer-Verlag, Berlin, pp. 41–60.

Pahl, G. and Beitz, W. (1996), *Engineering Design*, Springer-Verlag, Berlin.

Roy, B. (1991), The outranking approach and the foundations of ELECTRE methods, *Theory Decision*, 31(1), 49–73.

Roy, B. (1996), *Multicriteria Methodology for Decision Aiding*, Kluwer Academic, Dordrecht.

Saaty, T.L. (1990), *Multicriteria Decision Making: The Analytic Hierarchy Process*, 2nd ed., RWS Publications, Pittsburgh.

Sullivan, L. (1986), Quality function deployment, *Qual. Prog.*, 19(6), 39–50.

Urban, G.L. and Hauser, J.R. (1993), *Design and Marketing of New Products*, Prentice Hall International, Englewood Cliffs, NJ.

Vansnick, J.C. (1986), On the problem of weights in multiple criteria decision making (the noncompensatory approach), *Eur. J. Operational Res.*, 24, 288–294.

Vincke, P. (1992), *Multicriteria Decision Aid*, John Wiley & Sons, Chichester.

Wasserman, G.S. (1993), On how to prioritize design requirements during the QFD planning process, *IIE Trans.*, 25(3), 59–65.

9 How to Improve the Use of Quality Function Deployment

9.1 INTRODUCTION

In the previous chapters, we have seen that quality function deployment (QFD) is certainly an innovative tool for the development of a new product and many enterprises have decided to use it in such a way to improve their design cycle [Akao, 1990; ASI, 1987; Urban and Hauser, 1993]. Nevertheless, the obtained results have fallen short of expectations.

There are many causes of this only *partial* outcome [Cohen, 1995; Zairi and Youssef, 1995]. Besides natural suspiciousness toward the use of new methodologies, the main highlighted problems are:

- Cultural barriers that thwart the creation of project teams able to use QFD
- Lack of friendly tools able to reduce the training time
- Exponential growth of managerial difficulties connected with the increased size of design projects

Although the tool has been brilliantly employed in many applications [Sullivan, 1986; Akao, 1992; Griffin and Hauser, 1992; Franceschini and Rossetto, 1995a; Glushkovsky et al., 1995], giving a definitive *shape* to the concept of *customer-oriented design* [ASI, 1987; Hauser and Clausing, 1988; Franceschini and Rossetto, 1995b], in other situations it has given unsuitable responses.

By leaving organizational problems out of consideration, the real Achilles' heel of QFD is the management of designs of large size, which involves both a high number of customer requirements and a high number of technical characteristics. A preliminary way to overcome this problem is the decomposition of a project into a set of subprojects. This solution, however, is only a palliative, because it does not solve the root problem of managing and analyzing large relationship matrices.

A solution could be that of introducing some tools able to automate some activities that are carried out manually today, and to simplify the analysis of information contained in the house of quality (HoQ). Tools of this kind are, for example, methods for *clustering* technical characteristics [Kihara, Hutchinson, and Dimancescu, 1994], computerized methods, *group decision support system* (GDSS), *quantitative* engineering analysis techniques [Maier, 1995], and methods for setting up the design with reference to the competition [Franceschini and Rossetto, 1998, see also Chapter 6].

In addition to this summary list, two other methods are proposed in this chapter, the first directed toward the simplification of correlation matrix building and the second, to the determination of the minimum set covering of characteristics able to globally answer all customer requirements.

9.2 HOUSE OF QUALITY SUPPORTING TOOLS

As observed previously in Section 9.1, a large size HoQ inevitably tends to attenuate the *visibility* and facilities of analysis of the information contained there. To avoid a consequent debasement of advantages deriving from QFD, two further methods are proposed, in such a way as:

- To make easier the *compilation* of the correlation matrix among technical characteristics or among customer requirements
- To automatically verify the presence of technical characteristics or customer needs not related to other requirements or technical characteristics in the relationship matrix, respectively
- To identify the minimum set covering of technical characteristics able to cover all customer requirements

9.2.1 Method to Support the Compilation of the Correlation Matrix

The compilation of the *correlation matrix* can entail a big waste of time if the project manages many technical characteristics. To make this activity easier, one can think of automating the procedure, empowering the designer to establish only the *sign* of correlation.

To better understand the proposed method, it is important to define the meaning of *correlation* and *sign of correlation*.

In QFD, two technical characteristics are defined to be correlated if variations on the first one determine variations on the second and vice versa. The sign defines the direction of such a correlation (*positive*, if positive variations of the first are connected with positive variations of the other; *negative*, if the opposite). A significant example about the meaning of correlation and sign of correlation can be found in Hauser and Clausing (1988).

At present, designers who use QFD establish correlations among characteristics on the basis of *merely qualitative reasonings* [Cohen, 1995; Wasserman, 1993]. They do not take into account how technical characteristics influence customer requirements by means of the "content" of relationship matrix $\mathbf{R} \in \Re^{m,n}$ (m is the number of customer requirements and n is the number of technical characteristics).

By observing a generic relationship matrix, it may be noted that in many cases correlated characteristics influence the *same* customer requirements. This remark can be used as a starting point to build a partially automatic tool to define indirectly correlations among technical characteristics (to be used together with qualitative analysis).

As a matter of fact, if the i-th characteristic influences u-th, v-th, etc. requirements, it is likely that the j-th characteristic correlated to it influences the same requirements. Moreover, if the *dependence* among characteristics *induced* by the *action* of the same requirements may imply the presence of a correlation, the opposite it is not necessarily true. In fact, it can be demonstrated that a correlation between two characteristics may exist, without *induced links* on requirements in the relationship matrix.

Consequently, the method proposed here, which investigates the *induced dependence*, can highlight only a fraction of the total correlations. The presence of an induced dependence on the requirements is, therefore, a necessary but not sufficient condition to state that two characteristics are correlated. It is the designer, playing a new role of "validator," who must confirm the possible sufficiency.

To formulate the existence of dependencies induced by requirements an n-dimensional space constituted by a set of column vectors $\mathbf{b}_i \in \Re^n$ is considered (each one associated to a well-defined technical characteristic in the relationship matrix). Supposing that the relationship matrix $\boldsymbol{R}$ is filled adopting the symbol ● to individuate strong relationships, the symbol ◐ for medium relationships, and the symbol ○ for weak relationships, the coefficients of vectors $\mathbf{b}_i$ ($\forall i = 1, \ldots, n$) are determined as in the following:

$$\forall i,j \text{ if } r_{ij} = \text{●} \text{ or } \text{◐} \text{ or } \text{○}$$

then

$$b_{ij} = 1$$

Thus, by starting from the symbolic matrix $\mathbf{R}$ a new binary matrix $\mathbf{B} \in \Re^{m,n}$ is created.* Matrix $\mathbf{B}$ columns are then normalized producing another matrix $\mathbf{N} \in \Re^{m,n}$, with the columns named v_i ($\forall i = 1, \ldots, n$). The example in Figure 9.1 can give a better idea of the building process of matrix $\mathbf{N}$.

To represent the effects of the interdependence between i-th and j-th characteristics, the coefficient q_{ij} (scalar product of vectors $\mathbf{v}_i$) is introduced:

$$q_{ij} = \mathbf{v}_i^T \cdot \mathbf{v}_j = \cos\left(\mathbf{v}_i, \mathbf{v}_j\right) \qquad \forall i,j = 1, \ldots, n$$

By calculating q_{ij} for all pairs of vectors in the N matrix, it is possible to determine the *characteristics dependence matrix* $\mathbf{Q}$:

$$\mathbf{Q} = \mathbf{N}^T\mathbf{N}$$

$\mathbf{Q} \in \Re^{n,n}$ is symmetrical, with $q_{ii} = 1$; $\forall i = 1, \ldots, n$.

* Alternatively to the proposed procedure, matrix $\mathbf{B}$ could be defined as making a distinction between strong, medium, and weak relationships.

$$\mathbf{R} = \begin{bmatrix} \bullet & ◐ & \\ ○ & & ◐ \\ & ○ & \end{bmatrix}$$

$$\mathbf{B} = \begin{bmatrix} 1 & 1 & 0 \\ 1 & 0 & 1 \\ 0 & 1 & 0 \end{bmatrix}$$

$$\mathbf{N} = \begin{bmatrix} \frac{1}{\sqrt{2}} & \frac{1}{\sqrt{2}} & 0 \\ \frac{1}{\sqrt{2}} & 0 & 1 \\ 0 & \frac{1}{\sqrt{2}} & 0 \end{bmatrix}$$

FIGURE 9.1 Example of the building process of matrix **N** starting from matrix **R**.

Matrix **Q** expresses the degree of induced dependence among technical characteristics with reference to their capacity of influencing the same customer requirements.

It may be observed that, if one works with large matrices, the determination of matrix **Q** reveals the existence of columns or rows without relation with other columns or rows of the matrix **R**, respectively. This fact is highlighted by appearance of some *zeros* in the main diagonal of **Q**.

To allow the *filling* of the roof of the HoQ, information contained in **Q** is compared with a prefixed threshold k (with $0 \le k \le 1$); $\forall i, j$ if $q_{ij} > k$ then a *potential* correlation between characteristics i-th and j-th is admitted, else this correlation is supposed nonexistent. So starting from **Q**, a new matrix "roof" $\hat{\mathbf{Q}}$ is built.

At this point, the designer, on the basis of $\hat{\mathbf{Q}}$ coefficients establishes, in an interactive manner, the real existence of correlations (sufficiency condition) and its sign.

9.2.2 Minimum Set Covering of Technical Characteristics

In some situations it is important to define the *minimum set* of characteristics able to interact with all customer requirements.

The definition of priorities as carried out according to the traditional *scoring method* [Akao, 1990] presents some counterindication because it does not take into account *dependencies* among characteristics themselves. In fact, in many applications it happens that in the first ranking positions only strongly dependent characteristics appear, thus influencing only a limited part of customer requirements.

In some situations, however, it may be important for the designer to establish the minimum set of characteristics able to focus the *design attention* toward the main characteristics of a project development. Obviously, this does not mean that some characteristics can be neglected during the process development of a new product, but that the project can be organized in such a way as to give more importance to those characteristics that have more impact with customer requirements.

The search for the minimum set of technical characteristics covering all customer requirements is a classic combinatorial optimization problem known as the *set covering problem* [Nemhauser and Wolsey, 1988; Parker and Rardin, 1988].

In more detail, if $M = \{1, \ldots, m\}$ is a finite set and $\{M_j\}$, for $j \in N = \{1, \ldots, n\}$, a given collection of subsets of M, we say that $F \subseteq N$ covers M if $\bigcup_{j \in F} M_j = M$. The sets M_j are known as *covering sets*. If p_j is the cost associated to each M_j, the *minimum set covering* problem becomes that of *minimum-cost set covering*.

The search for the minimum number of columns (technical characteristics) able to cover all rows (customer requirements) is a *set covering problem* with $p_j = 1$, $\forall j \in N$. The set covering problem belongs to the set of *NP-complete* problems, which have a nonpolynomial computational complexity [Parker and Rardin, 1988].

In our specific case, since the aim is to give an *agile* supporting tool to the designer in the management of large relationship matrices **R**, a heuristic algorithm [Nemhauser and Wolsey, 1988] has been utilized (see Appendix). The algorithm has a polynomial computational complexity, and it is particularly suitable to give quick responses in a short time.

9.3 APPLICATION EXAMPLE

An example can make the proposed methods clearer. Figure 9.2 represents the HoQ for a design of a new undergraduate *curriculum* in the Mechanical Engineering Department at the University of Wisconsin-Madison [Ermer, 1995]. It contains 28 customer requirements and 17 technical characteristics of the service.

Actors in the design process have been students (as customers) and a team of department members (as designers). The design of the curriculum has been carried out by means of QFD.

To each customer requirement a degree of importance is assigned. A special score is also associated to each technical design characteristic of the matrix to determine a prioritization order. According to the *scoring method*, the technical importance rating w_j is given by the weighted column sum of each customer requirement by the *quantified* relationship values of technical characteristic i-th, formed by substituting 5 points for strongly related requirements, 3 points for medium related requirements, and 1 point for weakly related requirements (see Chapter 4, Section 4.6).

The scoring method applied to obtain the prioritization of technical characteristics is independent from the proposed compilation procedure of the correlation matrix. Even in this example, with a 28 × 17 matrix ($\mathbf{R} \in \Re^{28,17}$), the compilation of the correlation matrix associated to technical characteristics becomes very expensive.

By applying the method to **R** we are able to determine the *dependence* matrix $\mathbf{Q} = \mathbf{N}^T\mathbf{N}$. The obtained matrix **Q** is illustrated in Table 9.1. The generic coefficient q_{ij} of the matrix **Q** gives an indication of the degree of dependence between characteristics i-th and j-th.

TABLE 9.1
Correlation Matrix Q Associated to the Relationship Matrix R of Figure 9.2

	C1	C2	C3	C4	C5	C6	C7	C8	C9	C10	C11	C12	C13	C14	C15	C16	C17
C1	1	0	0	0.17	0.13	0.11	0,46	0	0	0.48	0.62	0.8	0	0.17	0.14	0	0
C2	0	1	0.58	0.32	0	0.2	0	0.29	0	0.45	0	0	0.29	0.32	0	0	0
C3	0	0.58	1	0.73	0	0.24	0	0.17	0	0.26	0	0	0.33	0.37	0	0	0
C4	0.17	0.32	0.73	1	0.16	0.13	0	0	0	0.14	0	0	0.37	0.6	0	0	0
C5	0.13	0	0	0.16	1	0.61	0.14	0.29	0.41	0.11	0.14	0	0.43	0.16	0.53	0.29	0.5
C6	0.11	0.2	0.24	0.13	0.61	1	0.35	0.47	0.33	0.09	0.12	0.2	0.47	0.13	0.44	0.35	0.41
C7	0.46	0	0	0	0.14	0.35	1	0	0	0.26	0.5	0.72	0	0.18	0	0	0
C8	0	0.29	0.17	0	0.29	0.47	0	1	0.47	0.13	0	0	0	0	0.46	0.83	0.58
C9	0	0	0	0	0.41	0.33	0	0.47	1	0.18	0	0	0	0.26	0.65	0.47	0.82
C10	0.48	0.45	0.26	0.14	0.11	0.09	0.26	0.13	0.18	1	0.52	0.56	0.26	0.42	0.24	0.13	0
C11	0.62	0	0	0	0.14	0.12	0.5	0	0	0.52	1	0.72	0	0.18	0.31	0	0
C12	0.8	0	0	0	0	0.2	0.72	0	0	0.56	0.72	1	0	0.16	0.13	0	0
C13	0	0.29	0.33	0.37	0.43	0.47	0	0	0	0.26	0	0	1	0.37	0.15	0	0
C14	0.17	0.32	0.37	0.6	0.16	0.13	0.18	0	0.26	0.42	0.18	0.16	0.37	1	0.17	0	0
C15	0.14	0	0	0	0.53	0.44	0	0.46	0.65	0.24	0.31	0.13	0.15	0.17	1	0.46	0.53
C16	0	0	0	0	0.29	0.35	0	0.83	0.47	0.13	0	0	0	0	0.46	1	0.58
C17	0	0	0	0	0.5	0.41	0	0.58	0.82	0	0	0	0	0	0.53	0.58	1

As is confirmed by Table 9.1, matrix **Q** is effectively symmetrical. In Table 9.1 values with $q_{ij} > k$ are highlighted, having fixed in this specific case a threshold $k = 0.6$. Dependencies so determined are as follows (Figure 9.2):

$$\{c_1, c_{11}\}, \{c_1, c_{12}\}, \{c_3, c_4\}, \{c_4, c_{14}\}, \{c_5, c_6\}$$

$$\{c_7, c_{12}\}, \{c_8, c_{16}\}, \{c_9, c_{15}\}, \{c_9, c_{17}\}, \{c_{11}, c_{12}\}$$

where c_j (with $j = 1, 2, \ldots, 17$) are the technical characteristics of the service.

It is intuitive that increasing the k value determines a decrease in the number of dependencies among pairs of characteristics. Having identified the pairs of characteristics with a degree of induced dependence greater than k, it is up to the designer (team department members) to evaluate their real consistency, establishing the positive or negative sign of correlation. Thus, for example, in our specific case all dependencies are translated in correlations except for $\{c_9, c_{15}\}$, which is rejected because it does not have the necessary requirements.

In addition to the set of induced dependencies and to correlations confirmed by the designer, *semantic* correlations also exist, identified on the basis of qualitative reasonings and technical considerations (e.g., mechanical characteristics that are physically related). The two contributions are merged, allowing us to build the roof of the HoQ. From an operating point of view, at the beginning we apply the automatic proposed procedure; subsequently, designers give their contributions based on specific technical considerations.

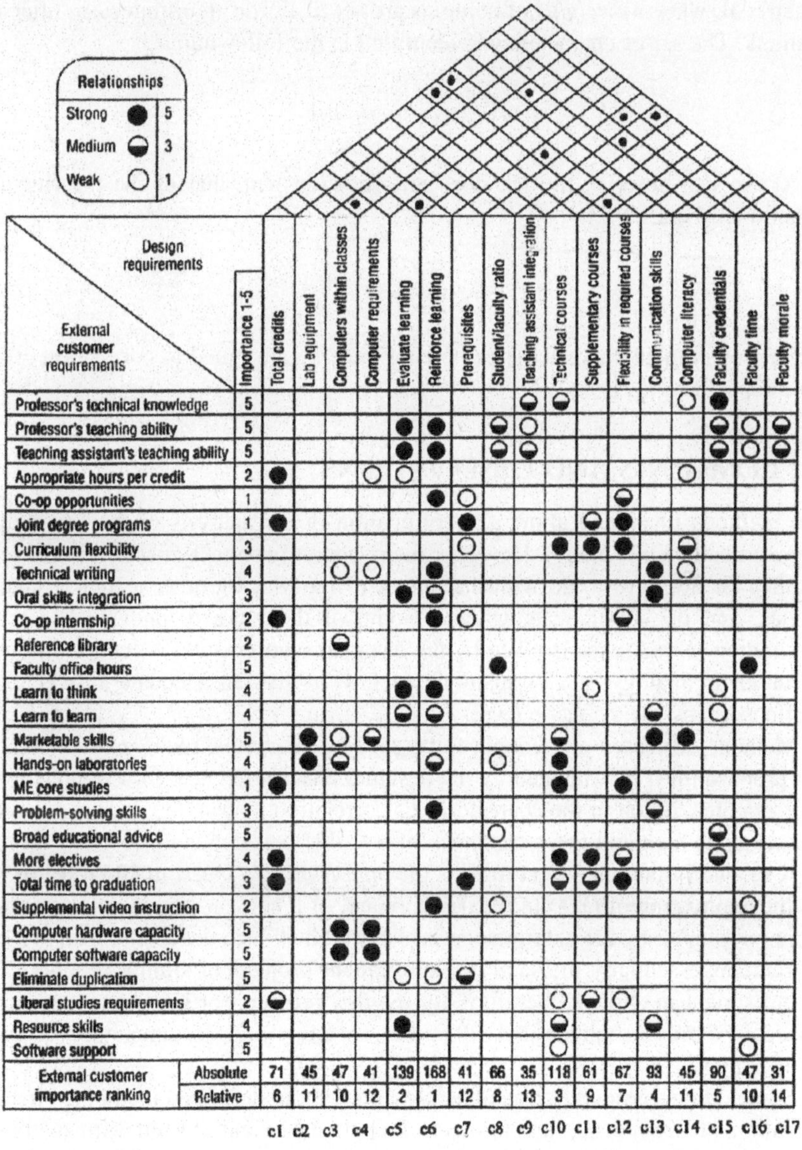

FIGURE 9.2 House of quality for the design of a new undergraduate curriculum c/o Mechanical Engineering Department at the University of Wisconsin-Madison. (From Ermer, D.S. [1995], *Qual. Prog.*, 28(5), 131–136. With permission.

It is still important to observe that the individuation of dependencies can be useful also for eliminating redundancies and duplications of characteristics expressed only with different terms.

At this point, it is possible to determine the minimum set covering of characteristics able to cover all customer requirements. Nemhauser's algorithm has been applied to matrix **B**. In this example, the set M coincides with the number of columns of matrix **B**, whereas *weights* of columns are set to 1. The algorithm stops after five iterations. The set of characteristics identified is the following:

$$\{c_6, c_{10}, c_3, c_1, c_8\}$$

As we can observe, the list does not coincide with that of the prioritization method (see Figure 9.2):

$$\{c_6, c_5, c_{10}, c_{13}, c_{15}\}$$

Differences in main characteristics ranking must be attributed to correlation effects (see the pair $\{c_5, c_6\}$).

9.4 COMMENTS AND OBSERVATIONS

Both methods illustrated allow a simplification in the analysis of the information contained in the QFD matrix. However, their support is only "syntactic," in the sense that they do not have to do with the matter of introducing other technical characteristics, and do not indicate whether some of them are redundant. The major contribution in this sense is given by the designer who operates — continuing with the linguistic analogy, at a "semantic" level — to give an adequate answer to the questions set by the syntactic level.

Methods discussed are thought to be used as interactive tools, and the solution they propose must be validated by the designer. With reference, for example, to the *automatic* generation of the characteristics correlation matrix, the designer is called to express the sign and the consistency of possible correlations.

A second important aspect regards the *opportunity to invest* in QFD. If it is true that the application of QFD in the development of a new project calls for a considerable investment by the enterprise, it is also true that once the design path is traced this remains essentially the same for subsequent projects of similar products. The big investment in QFD is, therefore, for the *first* project. For subsequent projects it is necessary only to update and adapt the procedure and documentation arranged the first time.

Some other considerations can be presented about described methods. First, the evaluation of dependence among vectors also can be carried out with nonbinary matrices, obtained, for example, by substituting the score determined with the traditional scheme of coding. Second, it can be observed that the evaluation scheme of dependences also can be applied for evaluating possible correlations among customer requirements. In this case correlations can be interpreted as an indication of the degree of overlapping of customer requirements.

Finally, it is still important to note that the application of a heuristic set covering algorithm can be extended to the relationship matrix with weighted columns, and that, in general, the obtained solution is not unique.

9.5 CONCLUSIONS

This chapter has presented a general discussion about the methodologies able to facilitate the use of QFD. Two methods able to simplify the analysis of the information contained in the HoQ have been proposed. The main aims of such methods are to reduce project development time and to alleviate designers from a series of activities that can be easily automated.

Appendix — Nemhauser's Heuristic Algorithm [Nemhauser and Wolsey, 1988]

INITIALIZATION

$$M_1 = M,\ N_1 = N,\ t = 1$$

SOLUTION GENERATION AT STEP $t > 1$

(a) Select $j^t \in N$: $\min\{p_j/\max|M_j \cap M^t|\}$

$$N^{t+1} = N^t \backslash \{j^t\}$$

$$M^{t+1} = M^t \backslash M_j$$

If

$$M^{t+1} \neq \varnothing$$

then $t = t + 1$, go to step (a).
If

$$M^{t+1} = \varnothing$$

then STOP.
The final greedy solution is given by all elements $j \notin N^{t+1}$, whereas its total weight, or cost, is given by $\sum_{j \notin N^{t+1}} p_j$.

REFERENCES

Akao, Y. (1990), *QFD: Integrating Customer Requirements into Product Design*, Productivity Press, Cambridge, MA.

Akao, Y. (1992), Origins and Growth of QFD, First European Conference on Quality Function Deployment, Milano.

ASI (1987), Quality Function Deployment, Executive Briefing, American Supplier Institute, Dearborn, MI.

Cohen, L. (1995), *Quality Function Deployment: How to Make QFD Work for You*, Addison-Wesley, Reading, MA.

Ermer, D.S. (1995), Using QFD becomes an educational experience for students and faculty, *Qual. Prog.*, 28(5), 131–136.

Franceschini, F. (1993), Impostazione di progetti di grande dimensione: il vincolo della Qualità, *Logistica Manage.*, 36, 34–42.

Franceschini, F. and Rossetto, S. (1995a), Quality & Innovation: a conceptual model of their interaction, *Total Qual. Manage.*, 6(3), 221–229.

Franceschini, F. and Rossetto, S. (1995b), QFD: the problem of comparing technical/engineering design requirements, *Res. Eng. Design*, 7, 270–278.

Franceschini, F. and Rossetto, S. (1997), Design for quality: selecting product's technical features, *Qual. Eng.*, 9(4), 681–688.

Franceschini, F. and Rossetto, S. (1998), QFD: how to improve its use, *Total Qual. Manage.*, 9(6), 491–500.

Glushkovsky, E.A., Florescu, R.A., Hershkovits, A., and Sipper, D. (1995), Avoid a flop: use QFD with questionnaires, *Qual. Prog.*, 28(6), 57–62.

Griffin, A. and Hauser, J. (1992), Patterns of communication among marketing, engineering and manufacturing — a comparison between two new product teams, *Manage. Sci.*, 38(3), 360–373.

Hauser, J. and Clausing, D. (1988), The house of quality, *Harv. Bus. Rev.*, 66(3), 63–73.

Kihara, T., Hutchinson, C.E., and Dimancescu, D. (1994), Designing software to the voice of the customer: new uses of QFD and quantification method of type III for decomposition of the requirements, *Qual. Eng.*, 7(1), 113–137.

Maier, M.W. (1995), Quantitative engineering analysis with QFD, *Qual. Eng.*, 7(4), 733–746.

Nemhauser, G.L. and Wolsey, L.A. (1988), *Integer and Combinatorial Optimization*, John Wiley & Sons, New York.

Parker, R.G. and Rardin, R.L. (1988), *Discrete Optimization*, Academic Press, San Diego.

Sullivan, L., (1986), Quality function deployment, *Qual. Prog.*, 19(6), 39–50.

Urban, G.L. and Hauser, J.R. (1993), *Design and Marketing of New Products*, Prentice Hall International, Englewood Cliffs, NJ.

Voss, F. (1992), QFD: A Tool and a Process to Become a Time-Based Competitive Organization, First European Conference on Quality Function Deployment, Milano.

Wasserman, G.S. (1993), On how to prioritize design requirements during the QFD planning process, *IIE Trans.*, 25(3), 59–65.

Zairi, M. and Youssef, M.A. (1995), Quality function deployment: a main pillar for successful total quality management and product development, *Int. J. Qual. Reliability Manage.*, 12(6), 9–23.

10 Setting Up Sizable Projects – Constraints of Quality

10.1 INTRODUCTION

The fact that the competitiveness of a company depends on its ability to operate effectively with full respect to the quality concepts is a notion often reiterated in the foregoing chapters.

To achieve and maintain this ability the company must avail itself of an adequate and efficient organization. The term *organization* denotes a system of resources, instruments, and rules aimed at achieving relational fluidity between its various internal units and between these and the market (suppliers and customers).

Quality, as we have seen, has gradually permeated all aspects of company life [Feigenbaum, 1983]. The production units were the first to be involved and subsequently, one by one, all company structures, right up to the design groups.

Considering that the advantages deriving from the use of quality methods are often evident and measurable in every type of production process, they must be enormously superior when we deal with large-scale projects, if only for the scale factor involved.

When dealing with such projects, the preliminary planning of the work constitutes one of the most critical phases. Their ultimate success and any future developments depend on how well the initial activities are organized [Conti, 1989; Hill, 1992]. In other words, if project planning begins in a confused and badly structured manner, the design department will be forced to bear the negative consequences; and the path leading to the final objective will be found to be an uphill climb, right from the start. Hence, it is necessary to act in an organized manner, right from the start, so as to channel the various activities while limiting as much as possible any wastage of resources.

In the following pages we shall present in detail a real example, and describe the planning and development of a *programmable logic controller* (PLC) for a distributed control system executed using quality function deployment (QFD).

10.2 TRADITIONAL SETUP OF DESIGNS

The complexity of a project depends on a multiplicity of factors, such as the operative environment where it is executed, the resources available, the dimension of the product or service that is to produced, and so on. The one aspect that must be made clear from the very beginning is that the greater the complexity of a project, the more its planning and management need to be organized adequately.

Up until recently (and still found in many companies to this day), the planning of a project was left to the unhampered initiative of the *project leader.* The time needed for developing a project, how it was developed, and the design criteria applied were absolutely not codified. The company's traditional know-how within the sector in which it dealt constituted its one and only strength. Planning a project by analogy following the same steps and operating along the same lines, according to well-trodden itineraries, represented the standard conduct in design administration [Pugh, 1991].

The drawbacks deriving from such an attitude are innumerable. The project leader's personalized planning, the total lack of structure, the turnover of personnel (who do not find valid career motivations in teamwork), and the incorrectly estimated industrial risks are only a few of the principal difficulties encountered. Hence, it is necessary to introduce and use systematically those tools that from a point of view of quality will permit a design organization even in this extremely delicate preliminary phase. Also, the idea and the necessity, when dealing with sizable projects, is to utilize QFD as a tool for structural planning.

10.3 DESIGN OF A PROGRAMMABLE LOGIC CONTROLLER USING QUALITY FUNCTION DEPLOYMENT

The automation of processes has always played an important role in the field of production and in the services related to it. As automation spread, and with it the concept of integrated plant systems, the use of distributed control systems has become increasingly important. Usually these are constituted by a computer network, structured on multiple levels in hierarchical order, able to fully satisfy the need for supervision, control, and optimization within the plant and within the management of the production system.

The basic element in these architectures is the PLC, which is equipment able to perform all the lower ranking functions required for automating the processes and the auxiliary plants connected to them. PLC, designed as an industrial control system replacing relay panels or wired solid-state logic panels, can today perform more complex control functions. A PLC can easily be integrated in a computer integrated manufacturing system by means of a local area network (LAN).

The typical functions of a PLC are to supply readings of primary information about the plant (digital input signals, analog input measurements, digital output controls, analog output *set points,* counter readings, etc.) to execute settings, sequences, and logics; to communicate with intelligent instrumentation; and to communicate with the higher levels of the control architecture.

From the point of view of their construction, PLCs contain one or more microcomputers that interact among them and provide the synchronization of the input–output sections toward the field and the fulfillment of the described functions. The example we shall present effectively concerns the subproject of the automation apparatus at level 1 (PLC) for the distributed control system PRODAS [Franceschini and Mattiacci, 1990]. The planning, development, and execution of the PRODAS system have required an overall effort totaling about 300 man/year.

The customer for the product was well defined and the project took off with the precise objective of introducing, in an industrial context, rather antiquated, some innovative elements in managing and directing the plant.

Activity commenced immediately by determining customer *desiderata* (desires). The requisites, identified through a series of interviews involving the team project leaders and the customers, were the following:

- r_1 adaptability of most types of signals coming from the "field"
- r_2 reliability of communication among subsystems
- r_3 easy generation and maintenance of system database
- r_4 simple usage of the development and supporting tool
- r_5 simple man–machine interaction
- r_6 availability of interchangeable programming languages
- r_7 possibility of effecting functional segregation for significant sections in the plant
- r_8 simple operating procedures for testing, installation, maintenance, and inspection
- r_9 reduced times for development and setup of apparatus
- r_{10} possibility of acquiring remote radio transmitted information
- r_{11} possibility of telepowering the instrumentation on field
- r_{12} possibility of interfacing with "intelligent" instrumentation
- r_{13} easy availability and utilization of hardware and software components on the market
- r_{14} presence of the functions of time-tagging and report by exception
- r_{15} accurate diagnosis system
- r_{16} system expandability (modular dimensioning)
- r_{17} security access points for personnel having different tasks
- r_{18} reduced times for personalizing and programming the apparatus (flexible configuration)
- r_{19} compatibility of response times with the dynamics of controlled processes
- r_{20} autonomous functioning even in the absence of higher hierarchical levels (level 2)

The requirements thus determined were confronted with the product technical characteristics:

- p_1 type of mechanics
- p_2 voltage
- p_3 central processing unit (CPU) (multi-micro-architecture)
- p_4 mass memory
- p_5 communication protocols
- p_6 process and instrumentation interfacing
- p_7 field input–output cards

- p_8 programming languages
- p_9 standard hardware and software development tools
- p_{10} hardware and software diagnostics
- p_{11} concurrent program management
- p_{12} man–machine interface (MMI)
- p_{13} system database
- p_{14} system modular organization and expandability
- p_{15} supporting documentation
- p_{16} ability to implement logics, sequences, and loops
- p_{17} redundant configuration for the most significant elements of the equipment

The interactions among the characteristics and the requirements are shown in Figure 10.1. It illustrates those interactions having mild or weak relationships as well as those having strong relationships. It is worth noting that in this particular case the types of interactions identified between customer requirements and product characteristics are actually only two. The interactions among the product characteristics are also shown on the roof of Figure 10.1.

By using the method of *independent scoring*, which assigns a score of 3 to the weak interactions and a score of 5 to the strong interactions (see Figure 10.1), we determine which is the most important design characteristic: system database (p_{13}). On the right of the figure is shown a schematic representation of how customer requirements are positioned with reference to a market leader company. The successive step entailed confronting the product characteristics ($p_1, \ldots, p_{17}$) with the characteristics of its single parts (*critical part characteristics*). (See also Chapter 3, Section 3.4.)

The total number of subsystems determined was 15. Of these, merely as an example, we show the part deployment matrix for the subsystem concerning the MMI function practiced by means of a portable terminal, for usage in *run-time* conditions (Figure 10.2).

In Figure 10.2, the characteristics (p_1, ..., p_{17}) are shown as rows and the functions so determined s_1, ..., s_{13} are shown as columns [Schiani, 1988]:

- s_1 *Log-in* allows controlled access to a work session through the introduction of a password associated to a specific professional qualification.
- s_2 *General menu* allows connections through various displays of MMI.
- s_3 *System functions* allow interventions on the system by means of designated keyboard commands.
- s_4 *Plant monitor* allows direct access to plant variables.
- s_5 *Selective monitor of variables* allows monitoring of variables that may be grouped according to opportune selection criteria.
- s_6 *System events log* shows system events and alarms.
- s_7 *Selective log* (limited only to specific variables) is shown for plant events and alarms.
- s_8 *General plant events and alarms log* shows system or plant events and alarms subdivided into categories.

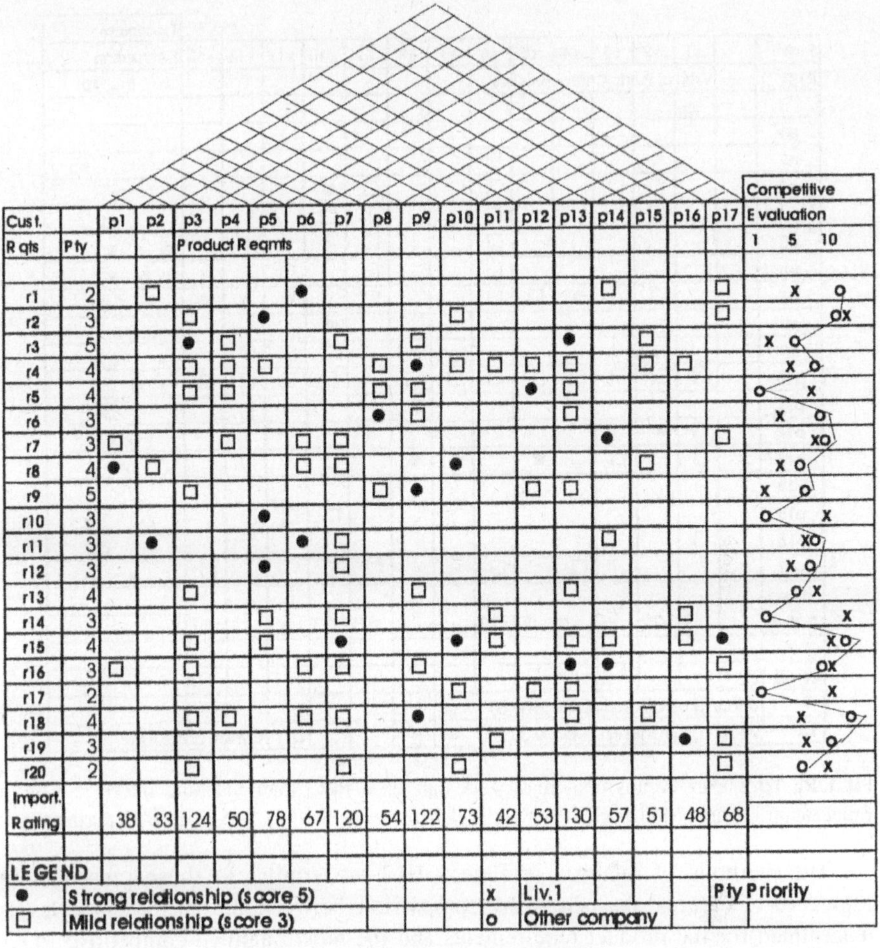

FIGURE 10.1 Product planning matrix for the design of a PLC. (From Franceschini, F. and Mattiacci, T. [1990], *L'elettrotecnica*, 77(10), 945–952. With permission.)

- s_9 *Alarms monitor* allows viewing a list of all the active alarms found in the plant or in the system.
- s_{10} *CPU diagnostics* are included.
- s_{11} *Input–output card diagnostics* are shown.
- s_{12} *Diagnostics* supplies the CPU diagnostics, I/O card diagnostics, and communication system diagnostics.
- s_{13} *Parameters displays* contain data characteristics of plant variables.

In short, we observe that the functions determined for the MMI run time subsystem on level 1 constitute only a limited set of those provided for the higher levels of PRODAS architecture (as a matter of fact, plant management at level 1 is required only in particular operational conditions).

Prdt		s1	s2	s3	s4	s5	s6	s7	s8	s9	s10	s11	s12	s13	Technical Evaluation
Rqts		Critical Parts Characteristics													1 5 10
p1															
p2							□								XO
p3			□	□			□			□	●				OX
p4			□				□				□				X O
p5							□			□			●		O X
p6							□			□		□			O X
p7			□		●	□	□			□		●		●	XO
p8											●				O X
p9											□	□	□		X O
p10		□	□	□	□	□	□	□	□		□	□	□	□	X O
p11															XO
p12		□	●												O X
p13		□	□	□	□	●		●	●		□	□	□		OX
p14															
p15											□	□	□		O X
p16		□													OX
p17					□		□			□	□			□	X O
Import. Rating		9	20	9	14	11	27	8	8	15	23	20	17	11	

LEGEND			
●	Strong relationship (score 5)	x	Liv.1
□	Mild relationship (score 3)	o	Other company

FIGURE 10.2 Part deployment matrix: design of a PLC man–machine interface. (From Franceschini, F. and Mattiacci, T. [1990], *L'elettrotecnica*, 77(10), 945–952. With permission.)

The methods of analysis in Figure 10.2 are similar to those presented in Figure 10.1. Figures also show the comparison between the performance levels determined for the product requirements and the most qualified competitive PLCs found on the market.

Proceeding in the analysis, the characteristics of the subsystem that has been identified are put in relation with the relative development process steps of each single function (*critical process steps*). Then module 3 is filled in (see Chapter 3, Figure 3.4). In this particular case, because we are dealing with a software product, the development modes follow the standard life cycle for software as foreseen by company norms [Telettra, 1988b; Telettra, 1990]. The cycle entails the following operative phases:

- Phase 1: software requirements review
- Phase 2: reliability evaluation
- Phase 3: functional analysis and project planning
- Phase 4: preliminary design and preliminary test planning
- Phase 5: detailed design and detailed test planning
- Phase 6: codification and *debugging*
- Phase 7: integration
- Phase 8: testing and validation

- Phase 9: release and filing of documents
- Phase 10: production, distribution, and installation
- Phase 11: maintenance and support

Within each phase, various activities of standard checks on the work carried out and on the relative state of advancement are foreseen (*design review*) [ESA–BSSC, 1991; Telettra, 1988b].

The other subsystems of the subplan (especially those involving hardware) are subjected to specific controls, detailed in the section on part process control parameters in module 3 (see Chapter 3, Figure 3.4).

In quite an analogous manner, with module 4 (the "process and quality control matrix," see Chapter 3, Figure 3.4) we define which are the quality standards imposed on the project plan and how they are to be implemented. In the case of the example at hand, the standards are shown in the *quality assurance plan* [ESA-BSSC, 1991, Rossi, 1985].

10.4 QUALITY FUNCTION DEPLOYMENT DEVELOPMENTS

The QFD tables allow us to obtain an integrated and structured overall view of the project, of the requisites to be satisfied, and of their relationships to the characteristics of the particular product or service. QFD maps, in particular, show profiles of the functions compared with those proffered by competitors (*technical evaluation*) limited to the upper limits of performance.

However, QFD tables do not explicitly present the various *alternatives* that can be used to "conceive" the single product functions p_i. In other words, although the QFD modules allow us to get a clear enough picture of the requisites and the functions with which the product or service must be endowed, no emphasis is placed on the alternative ways of carrying out a particular characteristic p_i. Using disparate strategies, of a technical nature as well as of an economic nature, one can, in point of fact, satisfy the totality of requirements expressed by the customer.

This observation suggests the introduction of a third dimension, perpendicular to the plane constituting customer and product requirements, able to determine alternative ways of attaining the various single product or service characteristics (see Figure 10.3).

In definition:

- $P = \{p_1, \ldots, p_m\}$ the set of technical product or service characteristics.
- $A_i = \{a_{i1}, \ldots, a_{in}\}$ the set of technological alternatives that could be chosen to fulfill the requirement p_i with $i = 1, \ldots, m$.
- $R_i = \{r_1, \ldots, r_k\}$ the set of criteria (coinciding with customer requirements), which intervene in the evaluation of the alternatives for the characteristic p_i with $i = 1, \ldots, m$ and $R_i \subseteq R$ (the set of criteria), each having its own evaluation scale (not necessarily numerical).

For each A_i, we encounter the problem of selecting a technological alternative a_i^* to best satisfy the preferences expressed on the criteria in R_i by the customer (decision maker).

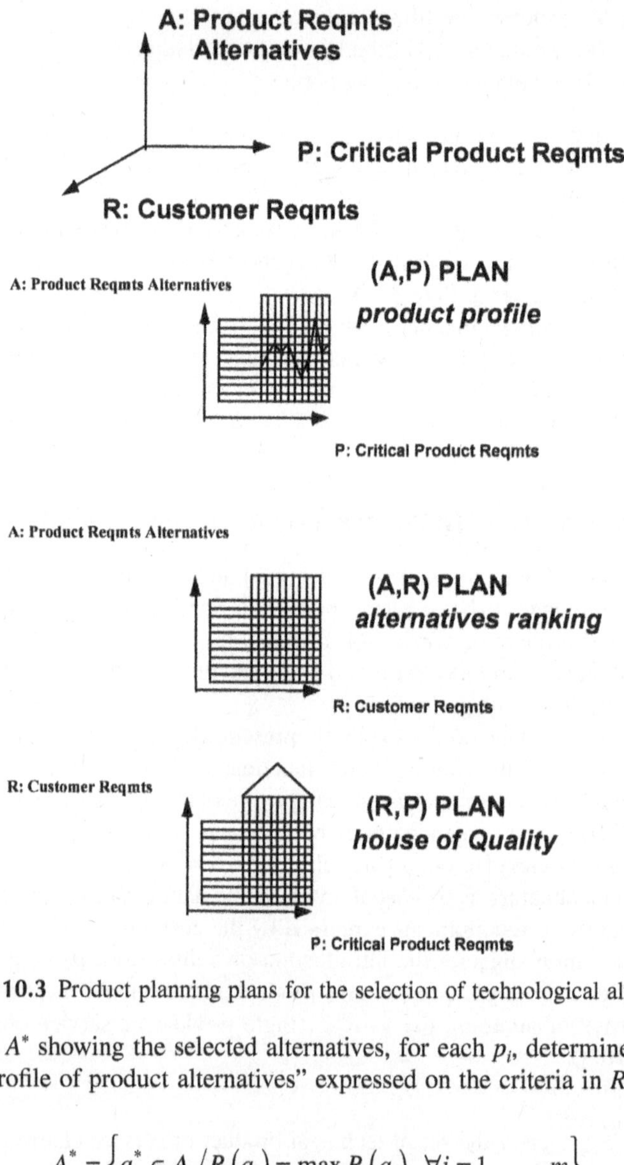

FIGURE 10.3 Product planning plans for the selection of technological alternatives.

The set A^* showing the selected alternatives, for each p_i, determines the "technological profile of product alternatives" expressed on the criteria in R.

$$A^* = \left\{ a_i^* \in A_i / R_i(a_i) = \max_{a_i \in A_i} R_i(a_i),\ \forall i = 1,\ \ldots,\ m \right\}$$

As is immediately evident, we once again face an MCDA problem (the maximum function is taken with respect to the system of customer preferences) [Ostanello, 1985; Roy, 1985].

It must be noted that the problem we have just mentioned, although making use of the same instruments utilized to determine the profile of a product compared with competitive items (see Chapter 6), is substantially different in nature from the latter.

TABLE 10.1
Selection of Technological Alternatives for the Programming Language Characteristic

Criteria	Alternatives a_1	a_2	a_3
r_5: man-machine interface (editor)	High	Medium	High
r_9: development and setup times (man/month)	57	53	75
r_4: usage of the software development tools (debugger, etc.)	Medium	Good	High
r_6: compatibility among different languages	Yes	Yes	No

The problem, in this case, is the identification of the most feasible technological solution able to carry out a certain technical performance.

From the point of view of the sequence of activities to be developed, we first determine the technical parameters to be assigned to the various technological alternatives identified when compared with the competition (see Chapter 6). At a successive stage we shall determine the most suitable technologies to be utilized.

An example may be useful to clarify the concept [Telettra, 1988a]. For the requisite p_8 (programming language), we are faced with the problem of choosing the type of memory format used to codify the programs (logics–sequences–loops) drawn up by the user. The languages examined were: list of instructions, functional blocks, and ladders, respectively.

The criteria that concurred in selecting the alternatives were (Table 10.1):

- r_5 MMI (*editor*)
- r_9 development and setup times
- r_4 simplicity of usage of the software development tools (*debugger*, etc.)
- r_6 compatibility among different languages

The technological alternatives identified were:

a_1) An ASCII format for the three languages
a_2) A compact format derived from the list of instructions for all three languages
a_3) A different format for the three languages

The evaluations of the technological alternatives available for the various criteria for requisite p_8 are shown in Table 10.1.

The analysis of each single product characteristic can be rapidly extended to all the other remaining characteristics. With reference to Figure 10.3, it is possible to identify three types of plans: plan (*A, P*) concerning product profile, plan (*A, R*) concerning the selection of alternatives, and plan (*R, P*) concerning the "canonical" house of quality (HoQ).

Each one of the plans we have determined is able to supply useful indications concerning the activities of design planning. In greater detail, plan (*A, P*) allows us to identify all the selected alternatives and to confront the various development strategies (contemporary evaluation of various alternative profiles). Plan (*A, R*) allows us to constantly check the evaluations of the alternatives, for each p_i, in relation to the different selection criteria.

REFERENCES

Baccalaro, W. and White, J. (1992), QFD and Quality in Design: Integrating QFD, DOE, Design Review and FMEA, First European Conference on Quality Function Deployment, Milano.

Conti, T. (1989), Process management and quality function deployment, *Qual. Prog.*, 22(5), 35–42.

Crow, K.A. (1992), Seminar on Concurrent Engineering, DRM Associates, Rome.

ESA–BSSC (1991), ESA Software Engineering Standards, ESA PSS-05-0, Issue 2.

Fiegenbaum, A.V. (1983), *Total Quality Control,* 3rd ed., McGraw-Hill Book, New York.

Franceschini, F. and Mattiacci, T. (1990), PRODAS (Process Optimization and Data Acquisition System): Un sistema per la gestione integrata dei servizi energetici, *L'elettrotecnica*, 77(10), 945–952.

Hill, A. (1992), New Product Introduction through QFD in a Total Quality Environment, First European Conference on Quality Function Deployment, Milano.

Ostanello, A. (1985), Outranking Methods in *Multiple Criteria Decision Methods and Applications*, Fandl, G. and Spronk, J., Eds., Springer-Verlag, Berlin, pp. 41–60.

Pugh, S. (1991), *Total Design — Integrated Methods for Successful Product Engineering*, Addison-Wesley, New York.

Rossi, F. (1985), Appunti su Qualità e normativa per il SW, Telettra serie Qualità e Affidabilità.

Roy, B. (1985), Methodologie multicritere d'aide à la decision, Ed. Libraire Economica, Paris.

Schiani, E. (1988), Specifica Funzionale del sottosistema MMI (Man Machine Interface) per il livello 1 del progetto PRODAS.

Sullivan, L. (1986), Quality function deployment, *Qual. Prog.,* 19(6), pp. 39–50.

Telettra, SpA (1988a), Normative per la Qualita' del SW. Standard 20 - V1.1.

Telettra, SpA (1988b), Progetto PRODAS alternative di progetto.

Telettra, SpA (1990), La specificazione del SW, 1.0.

Zucchelli, F. (1992), La Qualità Totale e il QFD, First European Conference on Quality Function Deployment, Milano.

11 Designing and Measuring Quality in the Service Sector

11.1 INTRODUCTION

The purpose of this chapter is to analyze more closely the service sector, which is in some ways similar to the production sector but has some particular characteristics differentiating it.

In particular, we shall present and analyze various analogies with products, and investigate the major differences from them, with reference to customer perceived quality in the service sector.

11.2 PARTICULAR CHARACTERISTICS OF THE SERVICE SECTOR

The necessity to define and measure quality undoubtedly stems from the manufacturing sector. In this field, different ways of understanding and conceiving quality have been developed, as is evident in the numerous definitions that have been coined [Crosby, 1979; Garvin, 1983; ISO 9000, 1994].

Our knowledge about quality referred to products, however, is not sufficient for us to understand the quality of services. To this end it may be interesting to consider which are the peculiarities of services compared with products. The main differences we are able to identify follow:

1. *Intangibility of services.* As we may observe in concrete cases, a service is, by its very nature, something intangible. Although a product is an object having precise physical characteristics, a service is in fact intangible, a performance. For this very reason it is difficult to apply to services the manufacturing concept that regards quality as conformity to specifications.

   With most services it is impossible to establish easily which qualitative elements are to be observed, which measures are to be taken, and which checks and tests must be carried out prior to sales to guarantee quality. Because of the intangibility of service, it may be difficult for a company to understand how customers will perceive the quality of a service they receive [Zeithaml, 1981].

2. *Limited standardization.* A service is subject to limited standardization and quantification because it is not static, but is highly dynamic. In any case, there are some elements that allow measurement and standardization, whereas others allow standardization in terms of expected behavior even if they are not easily quantifiable in terms of indexes. Others still defy standardization, or may be measured only generally according to the user's perceptions.
3. *Context setting (or inseparability).* A service is produced or supplied in the same place and cannot be stored. Therefore, service quality cannot be developed in a plant and subsequently delivered to the consumer. For example, in the case of services involving a great deal of manual labor, quality occurs while the service itself is being executed, usually during contact between the customer and the company employee [Lehtinen and Lehtinen, 1982].

   This entails the absence of any possibility of a repeat performance of the service, once it has been rendered. The lack of quality is blamed directly on the consumer and there is no possibility of correcting it. The personnel, the environment, and the organization are all there for consumers to see; these factors play their part in the service rendered and in the consumers' evaluation.
4. *Heterogeneity.* The heterogeneity of a service is tied to the strong incidence of the human factor, strongly subjecting it to the influence of the conditions and the context each time it is rendered. The services may vary from one producer to another, from one customer to another, and from one day to the next. The coherence in the behavior of personnel in the service sector is difficult to guarantee, because *what the firm intends to deliver may be entirely different from what the consumer receives* [Booms and Bitner, 1981].
5. *Reliability of human resources.* This is a fundamental factor in the service sector and not only where the consequences of possible errors may prove to be beyond repair for the persons involved. Many services, for example, those proffered over the counter, are rendered by persons holding positions that are considered very modest, badly paid, hardly prized, or not gratifying. The quality perceived by the user is the result of exchanges occurring with these persons.

Knowledge of the specific characteristics of services allows us to design the components supporting the planning of a quality control system. In particular, it means that we must define and evaluate the measurable elements (through reference standards), observable elements (behavioral standards), and elements defying standardization (perceptions).

The context setting implies defining and evaluating the relational, organizational and temporal variables of a service.

The reliability of human resources, in terms of quality, means that we must prevent, control, and correct the errors committed by the resources; and highlight their positive aspects.

Heterogeneity in services is expressed wherever possible by fixing the relevant standards, by endeavoring to achieve personalization, and by pursuing flexibility.

11.3 QUALITY STATUS OF ART IN SERVICES

Until recently, quality in the service sector aroused a rather limited amount of interest and only lately has it received a strong impulse to emerge, due to the rediscovery of its core importance by customers [Lehtinen and Lehtinen, 1982; Gronroos, 1982; Lewis and Booms, 1983; Sasser, Olsen, and Wyckoff, 1978].

The principal trails being investigated point to an effort to understand and study the following themes:

- Evaluations concerning quality are not expressed exclusively on the basis of the outcome of a service, but also on the *process* of supply of that service.
- Consumers' opinions on the quality of a service stem from the comparison between their expectations and the effective performance of the service.

When consumers buy an object, they use various tangible elements to judge the quality: style, form, color, label, consistency, packing, suitability for use, etc. Where services are concerned, however, such elements are less numerous. They are mostly limited, as a matter of fact, to the equipment physically present on the premises of those supplying the service, to their equipment, and to their personnel.

In the absence of tangible elements on which to base their evaluations of quality, consumers inevitably turn to other elements. Some authors maintain that the price becomes the basic reference index of quality in the absence of other orientating data [McConnell, 1968; Olander, 1970; Zeithaml, 1981].

The construction of quality for a service and the fixing of relative control procedures must be carried out in synergy with the design of the production–performance system. The quality of a service coincides, in point of fact, with

> the quality of the overall result of the process [...], and is directly proportional to the satisfaction the customer attains from the performance in its totality [Eigler and Langeard, 1988].

Both research workers and managers operating in the field of service companies agree that the quality of services brings about a comparison between expectations and performances. The quality of a service is the measure of how well the service performed corresponds to customer expectations.

Following up this criterion, Gronroos (1982) developed a model according to which he maintains that consumers, when they express their evaluation of quality, compare the service they expect to receive with the service they actually receive.

Sasser, Olsen, and Wyckoff (1978) proposed three different dimensions in the evaluation of performances in services: *levels of material, facilities*, and *personnel*. Implicit in this trilogy is the idea that the quality does not only affect the final result of the service but also the way it is delivered. Gronroos maintains that two types of quality in the service sector exist: the *technical quality*, which regards what the customer effectively receives from the service; and the *functional quality*, which is connected with the way the service itself is supplied [Gronroos, 1982].

11.4 CONCEPTUAL MODEL OF SERVICE QUALITY

In an article written by Parasuraman, Zeithaml, and Berry [1985] the authors first delineated a conceptual model (from this point on called the *PZB model*) of service quality, sufficiently generalized to be able to find application in the most disparate categories of services. The model, even though it presents obvious operational limitations (we are, after all, talking about a conceptual model), allows us to make sufficiently in-depth inquiries into the process of how service quality is delivered.

In constructing their model, the authors utilized sources of information such as interviews with customers; interviews with focus groups; and in-depth interviews with executives of companies proffering services in four different sectors: retail banking services, credit cards, securities brokerage, and product repair and maintenance of electrical equipment.

To illustrate the PZB model it is necessary to supply some preliminary definitions on the measurements it utilizes.

11.4.1 Definitions

11.4.1.1 Expected Quality (Q_a)

At given time *t*, expected quality is defined as the set of *n* functions of quality expected by the customer (Fa_i) and of the *n* weights (w_i) allowing us to take into consideration the relative importance that each single function embodies in the expectations of the customer:

$$Q_a = Q_a(w_i, Fa_i)$$

$$i = 1, 2, \ldots, n$$

The factors concurring to form expected quality are essentially four:

1. Specificity and peculiarity of the demand (*customer exigencies*)
2. *Past experience* of the service and the behavioral characteristics of the company supplying it
3. Customer-to-customer *internal verbal communications* (word-of-mouth communications)
4. Supplier-to-customer *external communications* (quality proposed by marketing)

11.4.1.2 Hypothesized Quality (Q_{ar})

The hypothesized quality (or the *reference quality*) is determined by an extended analysis of the demand carried out, for example, during investigations on *customer satisfaction* or when compiling the demanded quality table in the first quality function deployment (QFD) "house." (See also Chapter 4.) It entails an interpretation by the service company of the quality expected by the customer:

$$Q_{ar} = Q_{ar}\left(d_i,\ Ca_i\right)$$

$$i = 1,\ 2,\ \ldots,\ n$$

The Ca_i factors represent the *n functions of the expected quality hypothesized* by the marketing experts of the company (these are the *customer requirements* found in the demanded quality table of QFD). The coefficients d_i represent the weights assigned to each of these by the deliverer or supplier of the service (which are those placed in the degree of importance of requirements column in QFD). (See also Sections 4.2.5 and 4.5.2 in Chapter 4.)

11.4.1.3 Planned Quality (Q_d)

The *planned quality*, or *standard quality*, is determined on the basis of the knowledge of customer requirements and of company planning.

Let Ec_j and Fv_{jk} be the *engineering characteristics* and their relative *technological–economic constraints*, respectively (limitations presented by the generalized constraint k over the engineering characteristic Ec_j able to satisfy one or more corresponding functions of imagined quality Ca_j). The engineering characteristics can, therefore, be expressed as a set of functions:

$$Q_d = Q_d\left\{Ec_j\left(Ca_j,\ Fv_{jk}\right)\right\}$$

$$j = 1,\ \ldots,\ m$$

$$k = 1,\ \ldots,\ r$$

where m represents the number of service engineering characteristics and r represents the number of functions of the technological–economic constraints.

11.4.1.4 Offered Quality (Q_r)

The offered quality (or *quality of conformance*) is constituted by the totality of the *n functions of quality realized* (Fr_i) by the service:

$$Q_r = Q_r\left\{Fr_i\right\}$$

$$i = 1,\ \ldots,\ n$$

Taking into consideration the high incidence of the human factor, planned and realized specifications must also take into account training so as to ensure an appropriate development of personnel's professional capacity and ability and their motivation to work as *customer-driven* personnel.

11.4.1.5 Marketing Quality (Q_m)

Marketing, or *communicated*, *quality*, as we have had occasion to see earlier, contributes to determining expected quality and identifies the quality characteristics proffered to the customer to promote the sale of the service. In most cases, the communicated quality is given as greater than that effectively delivered, because customers are artificially promised *qualitative extras* ($p_j > 0$). The fact that this actually happens influences negatively their opinion and, consequently, the image they form (and transmit in customer-to-customer communications) about the service company. Therefore, we define marketing quality as a function that depends on the quality delivered and on the qualitative extras promised to the customer for each of the quality functions delivered:

$$Q_m = f(Q_r, p_j) = f(p_j, Fr_i)$$

$$i = 1, \ldots, n$$

$$j = 1, \ldots, q$$

The qualitative extras may vary according to times and contexts and depend on various factors, ranging from company marketing policy to the connotations of the demand and the direct supplier-to-customer contact.

11.4.1.6 Perceived Quality (Q_p)

Through an implicit relationship, perceived quality depends on the comparison of the various functions of quality as *perceived* by the customer (Fp_i) and the corresponding functions of expected qualities (Fa_i) and their weight (w_i):

$$Q_p = Q_p(Fp_i, w_i, Fa_i)$$

$$i = 1, \ldots, n$$

Perceived quality is the result of a global evaluation process. It is based on the comparison between the expected and the received service. It influences both the *hard* and the *soft* service components — and is measured according to value scales that vary in time for a given subject [Negri, 1992].

The soft, or immaterial, components stem from the relational dynamics in the supplier–customer interface. They are of primary importance, because it is the quality of these moments of truth that determines the company image projected externally [Carlzon, 1990].

Because many of the quality characteristics evaluated by the customer may be measured quantitatively by the service company, the procedures ensuring service quality must foresee the definition of a system of indicators able to verify them, by monitoring the numerical values inherent in the selected reference measurements [ISO 9004/2, 1991].

11.4.2 PZB Model

As confirmed unequivocally during interviews conducted on various focus groups, the basic idea underlying the PZB model is that the key to guaranteeing a good quality service consists depends on satisfying or exceeding customer expectations. The service quality the customer notices depends on the entity and the direction of the difference between expected and perceived quality. This difference or gap (Δ total) in turn depends on the nature of four other gaps (Figure 11.1) that emerge in the planning–production–presentation–delivering processes of a service:

$$\Delta \text{Total} = Q_a - Q_p = f(GAP1, GAP2, GAP3, GAP4)$$

11.4.2.1 Gap 1 — Discrepancy between Expected and Hypothesized Quality

$$GAP1 = Q_a - Q_{ar}$$

Many of the perceptions expressed by executives about what consumers expect from a quality service correspond to the expectations expressed by customers in focus groups, even if some differences do occur between executives' perceptions and customers' expectations.

Substantially, managers in service companies do not always identify ahead of time the characteristics denoting what a customer maintains to be a good quality level, which prerogatives a service must present to answer customer exigencies, and which levels of performance are necessary in these characteristics for them to deliver service quality [Langeard et al., 1981; Parasuraman, Zeithaml, and Berry, 1985].

The differences between expected quality and hypothesized quality are due to:

- The inadequacy of market research and the inefficient utilization (during planning, programming, and industrialization of a service) of the results obtained during feedback by the field
- The company organization chart structured as excessively articulated, hierarchical in structure, and bureaucratic
- A restricted flow of *bottom-up* information, limiting at the management and intermediate levels their knowledge of the quality functions expected by customers

11.4.2.2 Gap 2 — Discrepancy between Hypothesized Quality and Planned Quality

$$GAP2 = Q_{ar} - Q_d$$

A recurring theme in interviews among management in all four companies offering services is the difficulty encountered in trying to satisfy or exceed consumer expectations. Managers point out the limits that stop them from supplying what

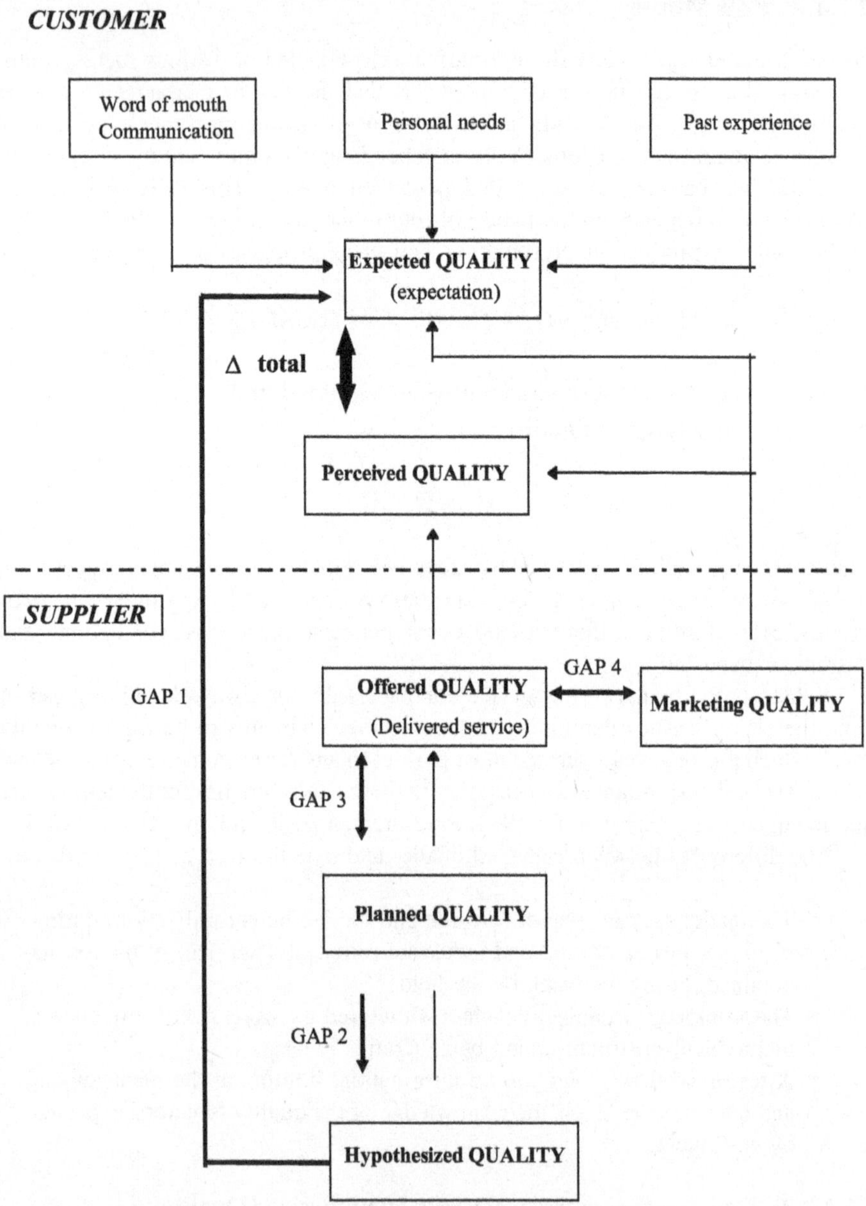

FIGURE 11.1 Service quality conceptual model. (From Parasuraman, A., Zeithaml, V.A., and Berry, L.L. [1985], *J. Mark.*, 49, 41–50. With permission.)

consumers expect. By way of example, managers of service companies proffering repairs know quite well that consumers consider the time taken to repair a failure an essential ingredient in high service quality. However, they have difficulty in establishing operational specifications that could render their services invariable and rapid, due to a lack of qualified personnel, and to the high degree of fluctuation encountered in the demand.

The discrepancy between hypothesized and planned quality, which ultimately does have repercussions on the perception of quality on the part of the consumer, is due to:

- An insufficient culture concerning service quality, which results in inadequate *quality plans* and in a marginal commitment by company management
- An orientation of company strategy toward internal organization and toward results (*bureaucratic dynamics* and *product-out*) instead of toward *process improvements* and *market-in*
- A presumable incompatibility among quality objectives and economic objectives
- A low level of industrialization in servicing and low standardization of process components or personnel's *tasks* [Levitt, 1992]

11.4.2.3 Gap 3 — Discrepancy between Planned and Offered Quality

$$GAP3 = Q_d - Q_r$$

Even where precise criteria for executing a service well and for correctly handling a customer do exist, there is no automatic certainty that the performance will be of good quality. Management in service companies knows that their personnel exerts a strong influence on the quality as it is perceived by customers, and that the personnel's performances cannot always be standardized. When asked which, according to them, is the principal cause of inconveniences in service quality, the agreed reply is: the role of personnel who come into direct contact with the customer.

To summarize, discrepancy between planned quality and effective quality is due to:

- Confused and approximate definition of the roles, duties, and objectives for quality
- Presence — to a greater or lesser degree — of personnel possessing inadequate capabilities or competence for their functions or for their position within the company structure
- Existence (which cannot be eliminated) of interface situations potentially exposed to a juxtaposition of roles between a service supplied externally (to the customer) and one applied internally (within the company)
- Poor or obsolete technology at the production or performance level and poor or obsolete aids to personnel's activities
- Inefficient control and evaluation systems [Peters and Waterman, 1984]
- Presence of decisional centering at the higher hierarchical levels, with rare and limited exceptions where decisional power is delegated to executive personnel [Carlzon, 1990]
- Lack of cohesion and group spirit and scarce aptitude to team work

11.4.2.4 Gap 4 — Discrepancy between Offered Quality and Marketing Quality

$$GAP4 = Q_r - Q_m$$

Advertising through the media and other company communications is able to influence the expectations of consumers. Where such expectations play an important role in the perception of service quality on the part of consumers, the company must be quite sure that it is not promising, through its communications, more than it will in fact be able to offer. This may result in an increase in initial expectations, but would then be followed by a lower perceived quality if the promises were not kept.

The discrepancy between offered quality and marketed quality is therefore the result of:

- Inadequate communications between the various company functions and within the single departments
- External communications (publicity, promotions, etc.) that tend to highlight the quality offered or to focalize only on the really excellent components, rendering the overall information quite unclear, misleading, and sometimes even inexact

11.5 SERVICE QUALITY DETERMINANTS

Focus groups have revealed that, irrespective of the type of service, consumers adopt fundamentally similar criteria to evaluate its quality. These criteria seem to fit into ten key categories, called "service quality determinants in the perception of service quality," and are described in Table 11.1 [Parasuraman, Zeithaml, and Berry, 1985]. For each determinant, the table supplies examples of specific criteria emerging from focus groups. It does not imply the complete independence of the ten determining factors.

The customer's viewpoint on service quality is shown in the upper part of Figure 11.1.

Figure 11.2 indicates that perceived service quality is the result of the comparison the consumer makes between the expected service and the service effectively received. It is quite probable that the relative importance of the ten determinants that bring about customer expectations (before service delivering) will later differ from their relative importance in the service effectively perceived by the consumer after its supply. It is important to note that the comparison between the service we expect and that effectively received is no different from the opinion we form when judging a product. The difference lies in the *nature* of the characteristics compared with which services are evaluated.

On the basis of the PZB model, it is possible to assert that perceived service quality is posited along a continuum ranging from ideal quality to totally unacceptable quality. The position of the perceived service quality on this continuum depends on the nature of the discrepancy between expected quality (Q_a) and perceived quality (Q_p):

TABLE 11.1
Determinants of Service Quality

1. **Reliability** involves consistency of performance and dependability; it means that the firm performs the service right the first time; it also means that the firm honors its promises; specifically, it involves:
   - Billing accurately
   - Keeping records correctly
   - Performing the service at the designed time
2. **Responsiveness** concerns the willingness or readiness of employees to provide service; it involves timeliness of service including:
   - Mailing a transaction slip immediately
   - Calling the customer back quickly
   - Giving prompt service (e.g., setting up appointments quickly)
3. **Competence** means possession of the required skills and knowledge to perform the service; it involves:
   - Knowledge and skill of the contact personnel
   - Knowledge and skill of operational support personnel
   - Research capability of the organization, e.g., securities brokerage firm
4. **Access** involves approachability and use of contact; it means:
   - Service easily accessible by telephone (lines not busy and callers not put on hold)
   - Waiting time to receive service (e.g., at a bank) not extensive
   - Convenient hours of operation
   - Convenient location of service facility
5. **Courtesy** involves politeness, respect, consideration, and friendliness of contact personnel (including receptionists, telephone operators, etc.); it includes:
   - Consideration for the consumer's property (e.g., no muddy shoes on the carpet)
   - Clean and neat appearance of public contact personnel
6. **Communication** means keeping customers informed in language they can understand and listening to them; it may mean that the company has to adjust its language for different consumers — increasing the level of sophistication with a well-educated customer and speaking simply and plainly with a novice; it involves:
   - Explaining the service itself
   - Explaining how much the service will cost
   - Explaining the trade-offs between service and cost
   - Assuring the consumer that the problem will be handled
7. **Credibility** involves trustworthiness, credibility, and honesty; it involves having the customer's best interests at heart; contributing to credibility are:
   - Company name
   - Company reputation
   - Personal characteristics of the contact personnel
   - Degree of hard sell involved in interactions with the customer
8. **Security** is the freedom from danger, risk, or doubt; it involves:
   - Physical safety (Will I get mugged at the automatic teller machine?)
   - Financial security (Does the company know where my stock certificate it?)
   - Confidentiality (Are my dealings with the company private?)
9. **Understanding/Knowing the customer** involves making the effort to understand the customer's needs; it involves:
   - Learning the customer's specific requirements
   - Providing individualized attention
   - Recognizing the regular customer

TABLE 11.1 (continued)
Determinants of Service Quality

10. **Tangibles** include the physical evidence of the service, such as:
    - Physical facilities
    - Appearance of personnel
    - Tools or equipment used to provide the service
    - Physical representation of the service, such as a plastic credit card or a bank statement

From Parasuraman, A., Zeithaml, V.A., and Berry, L.L. (1985), *J. Mark.*, 49, 41–50. With permission.

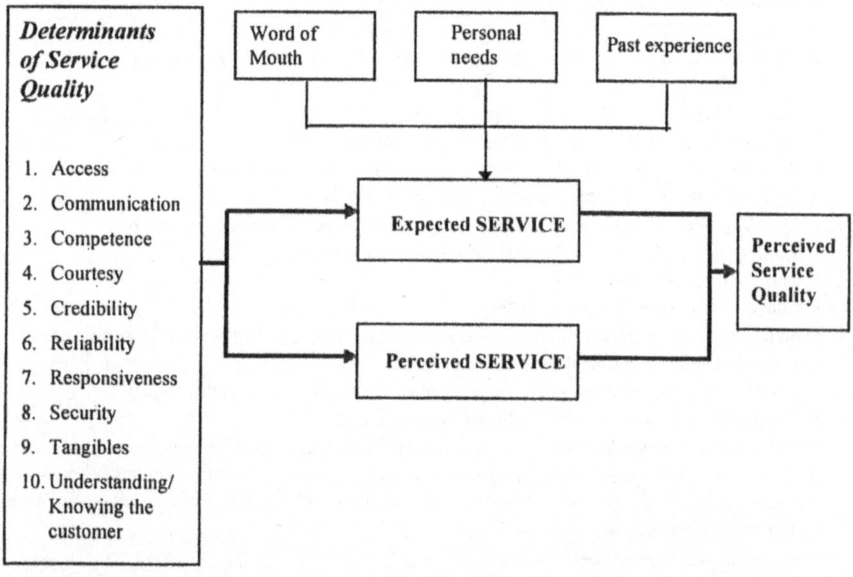

FIGURE 11.2 Determinants of perceived service quality. (From Parasuraman, A., Zeithaml, V.A., and Berry, L.L. [1985], *J. Mark.*, 49, 41–50. With permission.)

1. When $Q_a > Q_p$, perceived service quality is less than satisfactory.
2. When $Q_a = Q_p$, the perceived service quality is satisfactory.
3. When $Q_a < Q_p$, the perceived service quality is more than satisfactory.

Many authors have described numerous methods for evaluating service quality. Some are explanations of conceptual models designed to understand the mechanism of evaluation and include Parasuraman, Zeithaml, and Berry [1985, 1991]. Other methods are the result of empirical experimentation carried out on various applicative sectors.

11.6 QUALITOMETRO METHOD

The core difference between products and services is the great difficulty encountered in the latter to quantify them. This does not imply that products, considered during their whole life cycle (not only during the production step), entail a simple solution for the analogous problem of global quality quantification.

For most services it is not easy to define the qualitative elements that have to be observed, the measurements that should be taken, and — last but not least — the tests and checks that are needed to control them.

Many studies [Cronin and Taylor, 1992, 1994; Parasuraman, Zeithaml, and Berry, 1985, 1993] have strived to analyze how a customer perceives and evaluates the quality of received service. However, because the problem is complex, the results represent only a promising beginning. We offer a contribution to the growing discussion, and suggest a method for the evaluation and on-line control of the gap between expected and perceived quality. In this context, *on-line* means evaluating service during delivery, remembering a term from the control theory. The main innovations of this method are the operative procedures for data collection and data elaboration.

11.6.1 Problem of Quantifying Service Quality

Literature offers many different tools to evaluate service quality. SERVQUAL [Carman, 1990; Parasuraman, Zeithaml, and Berry, 1988, 1991, 1994] and SERVPERF [Cronin and Taylor, 1994; Parasuraman, Zeithaml, and Berry, 1994; Teas, 1994] are two of the most famous methods used as examples in the applications. Some interesting surveys can be found in Parasuraman, Zeithaml, and Berry, [1991] and Cronin and Taylor [1991].

Common features of all these methods are:

- The use of questionnaires
- The acknowledgment of the multidimensionality of quality (quality able to be expressed as a set of attributes)
- The method of considering expected and perceived quality or the latter only [Cronin and Taylor, 1992]
- The numerical interpretation of data collected in the questionnaires

Quantifying the attributes that define service quality is particularly difficult because quality is expressed not only by service outcome but also by its delivering process and by the contemporaneous presence of objective, subjective, relational, and organizational variables.

The problem of quantifying service quality finds its solution in first stating the following:

- Attributes [Garvin, 1987; Franceschini and Rossetto, 1995a] affecting service delivering and their relative importance for the customer
- Objective or subjective measurable elements
- Measuring systems able to evaluate attributes and variables
- Proper model to define the link between variables and delivering process
- Procedures needed to monitor service delivering continuously
- Reference standards

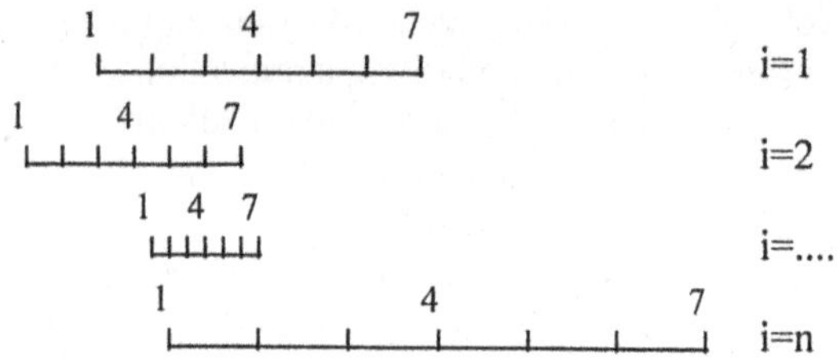

FIGURE 11.3 Reference systems for different individuals ($i = 1, 2, \ldots, n$) interviewed on a seven-point conventional scale. The figure shows the difficulty in aggregating collected data.

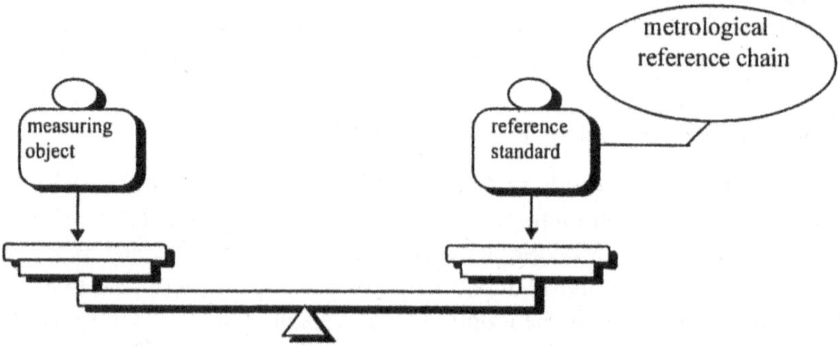

FIGURE 11.4 Conceptual pattern for an objective measurement.

People, mostly because of the diversity of their individual reference systems, influence service evaluations substantially. Figure 11.3 shows a possible situation for *n* reference systems disclosed by interviewees who showed their preferences on a conventional seven-point evaluation scale. Lining up and aggregating data collected from different individuals on conventional scales, as is usually the case on questionnaires, becomes difficult [Cronin and Taylor, 1994; Hayes, 1992; Parasuraman, Zeithaml, and Berry, 1988].

Figures 11.4 and 11.5 suggest a comparison between two conceptual models for an objective measure and a subjective evaluation. Both of them have the left scale pan for the object to be measured, the right scale pan for the "reference standard" for objective measurements, or the "comparison term" for the subjective ones. The reference standard constitutes one of the rings of the metrological reference chain. It begins with the national reference standard and continues with many intermediate standards, each characterized by an increased measurement uncertainty. Refer, for instance, to length measures: the reference standard chain begins with the sample meter and continues gradually up to the most common length measurement tools (graduated bars, cursor caliber, etc.).

For subjective measurements, no objective metrological reference chain exists, but a subjective reference system exists that is different for each individual. This aspect renders difficult the aggregation of data collected from different individuals.

Table 11.2 suggests a comparative pattern for objective measurement and subjective evaluation, referred to measurement tools and reference systems.

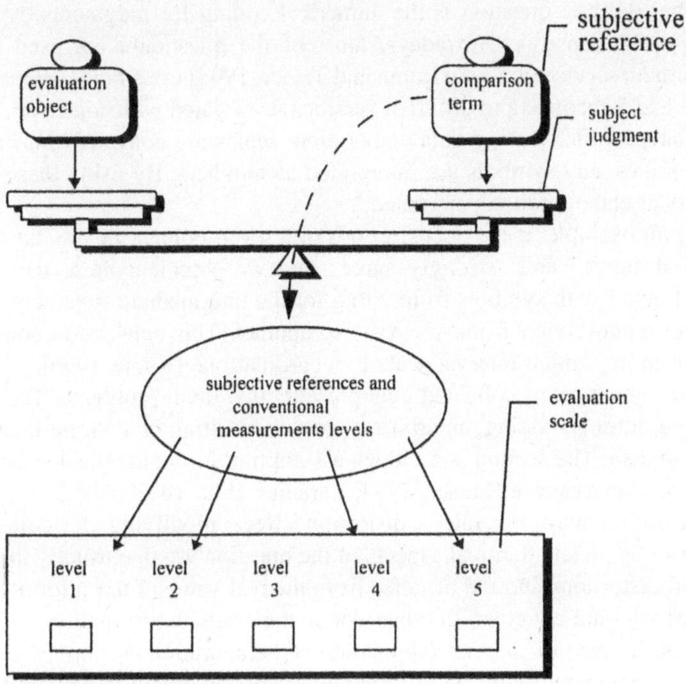

FIGURE 11.5 Conceptual pattern for a subjective measurement (evaluation).

TABLE 11.2
Comparative Pattern between Objective Measurement and Subjective Evaluation

	Objective Measurement	Subjective Evaluation
Measurement tools	Measurement tool	Judgment
Reference systems	Reference standard	Personal expectations

Thus, the homogeneity hypothesis adopted for individual reference systems is critical when applied to data aggregation and interpretation. Moreover, if we consider the effect of other hypotheses, suggested by other studies [Oliver, 1981], about the continuous perceived quality evolution in individuals who *periodically* use a service, we find another obstacle for the homogeneity hypothesis.

Many variations to the proposed methods have been introduced that try to "hang" reference systems. For instance, Parasuraman, Zeithaml, and Berry [1991, 1994] interviewed people, asking them (1) what they consider a high-quality enterprise and an inquiring company (called XYZ), (2) what they consider the scale end as the service delivered by high-quality enterprises and later how they consider XYZ, or (3) finally what they consider the scale medium level as the expected level for a high-quality enterprise and then XYZ according to anyone's expectations.

Another delicate question is the numerical coding for judgments expressed in interview questionnaires. Nowadays, most of the questionnaires used to collect information on service quality [Cronin and Taylor, 1994] use a point evaluation scale (1 to 7 or 1 to 5 according to different versions) associated with adjectives to qualify a particular position. During data elaboration, scales are converted into numerical interval* scales, and symbols are interpreted as numbers. By using these numbers, the statistical elaboration is performed.

Thus, for example, if the opposite ends of a seven-point scale are the statements "strongly disagree" and "strongly agree" and we associate these the numerical symbols 1 and 7 with symbols from 2 to 6 for the intermediate statements, we have carried out a conversion from one scale to another. This conversion constitutes in moving from an ordinal interval scale to a cardinal one [Rupil, 1996].

The scalarization of collected data presents two main problems. The first is in introducing, through coding, an arbitrary metric, resulting in a wrong interpretation of gathered data. The second is a hidden assumption for an identical scale interpretation by any interviewee [Fraser, 1994; Larichev et al. 1993, 1995].

Scalarization may generate a distortion effect, modifying the collected data partially or completely. Critical aspects of the question are that usually the entity of introduced distortions and the distance from the real value of the information given by the customer are not clear. In other words, the original information — arbitrarily *enriched* or directed to *simpler* aggregation and elaboration — may be excessively modified — compared with the information effectively expressed by customers — and the consequences are intuitable.

Another point to be considered is the *idiosyncratic effect* [Drew and Castrogiovanni, 1995] that encourages the interviewee to assume critical positions for the evaluation tool. The elaboration of a few intrusive and untiring tools for the customer is then an important issue.

These methods that simultaneously evaluate expected and perceived quality [Parasuraman, Zeithaml, and Berry, 1988, 1991; Teas, 1993, 1994] show a problem of mutual influence of the two variables. As a matter of fact, as the questionnaire is administered after service delivery, it may be expected that given judgments for expectations are *influenced* by the ones for perceptions and vice versa. This matter presents a further problem concerning the correct ways of delivering questionnaires.

11.6.2 Qualitometro Project

The *Qualitometro* method was created to evaluate and check on-line service quality. The tool was devised for executing various measurements of Q, Q_p and their differential δQ.

Evaluations of expected and perceived quality are carried out *separately* without danger of mutual influences. Expected quality is investigated *ex ante* whereas perceived quality sees *ex-post* service delivery. Figure 11.6 explains administration conditions.

---

* A linear interval scale allows object setting so that the differences between side elements are the same. Interval scales, without any scale origin, allow equality or unequality, ordering, and subtraction operations.

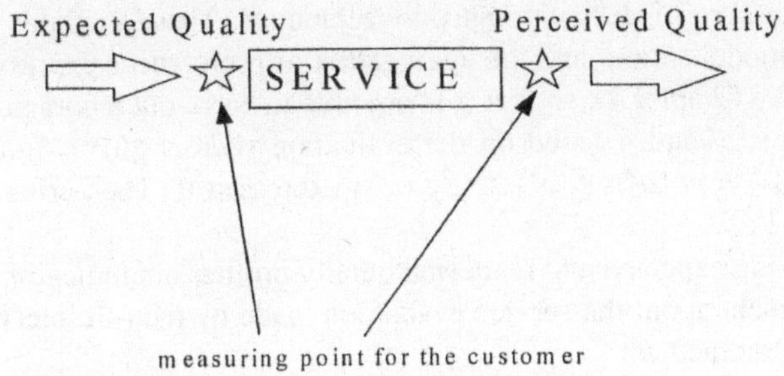

FIGURE 11.6 Administration pattern for Qualitometro questionnaire.

An interesting feature of this method is that the customer can estimate perceived service quality by directly comparing it with expected quality, before receiving it. In this way, customers are forced to express an evaluation using their own reference system.

The organization of the instrument, with a minimal variation, is based on the determinants of service quality shown in the PZB model [Parasuraman, Zeithaml, and Berry, 1985; 1993; 1996]. Unlike SERVQUAL or other similar instruments, however, Qualitometro is less intrusive with the customer. Every item is synthesized in a form, which shows the representative determinants of service quality, a few words of comment, and a seven-point scale for its evaluation.

Any point on the scale is representative of a particular state of the service defined by a proper adjective (see form instructions in the Appendix). In the questionnaire, an evaluation of the importance of the dimensions, with points between 1 and 7, is also required. A short introduction for using the form and an example are given.

The questionnaire pattern used in the Qualitometro project is shown in the Appendix. The form is for an experiment in progress at the Department of Systems of Production and Economics (DISPEA) library at Turin Polytechnic.

The extreme compactness of the instrument does not allow for verification of validity and reliability of data collected [Lehtinen and Lehtinen, 1982]. This operation can be performed only after the arbitrary cardinalization, as discussed.

Qualitometro allows two possibilities for data elaboration:

- Statistical data analysis according to a traditional procedure, after the scalarization of collected information
- Data analysis, without any artful numerical codification, based on the ordinal properties of collected information [Franceschini and Rossetto, 1998; Fraser, 1994]

A discussion on the latter possibility follows. The method begins by establishing, for each interviewee, the prevalence of the expected quality profile over the perceived quality profile ($Q_a > Q_p$), or vice versa. This activity is supported by one of the multicriteria decision analysis (MCDA) methods [Ostanello, 1985; Roy, 1991].

The ability to choose from various alternatives (expected and perceived quality profiles) compared with a set of evaluation criteria (qualitative or quantitative) constitutes the core of decision analysis techniques. These methods allow for the analysis, the modelization, and the aggregation of preferences expressed by a decision maker (see Chapter 7), so that it is possible to carry out a comparison between two alternatives a' and a based on the evaluation vectors $\mathbf{g}(a) = [g_1(a), g_2(a), \ldots, g_n(a)]$ and $\mathbf{g}(a') = [g_1(a'), g_2(a'), \ldots, g_n(a')]$ expressed for each criterion g_j where $j = 1, \ldots, n$.

By analyzing expected and perceived quality profiles, an indicator q_{ij} expressing a global judgment about the service evaluation made by the i-th interviewee for the j-th sample is defined as:

$$q_{ij} = \begin{cases} 1 \text{ if } Q_a \le Q_p \\ 0 \text{ if } Q_a > Q_p \end{cases}$$

$$\forall i = 1, \ldots, n$$

$$\forall j = 1, \ldots, k$$

Judgments made by various interviewee individuals are collected in an n-sized sample and described on a p control chart. The p-chart allows an interpretation during the time of the differential of service quality.

The percentage p_j for each sample is calculated as follows:

$$p_j = \frac{\sum_{i=1}^{n} q_{ij}}{n}$$

$$\forall j = 1, \ldots, k$$

where j is the sample number and k is the number of samples.

By assuming a constant size sample, the central line $\bar{p}$ and control chart limits are calculated as follows [Grant and Leavenworth, 1988]:

$$\bar{p} = \frac{\sum_{j=1}^{k} p_j}{k}$$

$$UCL = \bar{p} + 3\sqrt{\frac{\bar{p}(1-\bar{p})}{n}}$$

$$LCL = \bar{p} - 3\sqrt{\frac{\bar{p}(1-\bar{p})}{n}} \tag{11.1}$$

UCL and LCL are the upper and lower chart control limits, respectively. As is shown in Equation (11.1), the width between control limits is correlated with the sample size. With a p-chart, it is possible to carry out sampled, on-line service quality control.

Out-of-control points, or a particular collected data pattern, signal assignable cause variation in service delivery [Grant and Leavenworth, 1988; Hayes, 1992]. This unexpected event may be investigated in depth when examining single-dimension performance. For each dimension, a *p*-chart similar to the previous one is built in such a way as to quantify the d_{ijr} on a particular attribute:

$$d_{ijr} = \begin{cases} 1 \text{ if } r_a \leq r_p \\ 0 \text{ if } r_a > r_p \end{cases}$$

$$\forall i = 1, \; \ldots, \; n$$

$$\forall j = 1, \; \ldots, \; k$$

$$\forall r = 1, \; \ldots, \; z$$

where d_{ijr} is a quality index for *i*-th interviewee, for *j*-th sample, and for *r*-th dimension; r_a and r_p are the expected and perceived values, respectively, for *r*-th dimension; and *z* is the number of dimensions (determinants).

By examining the chart for the global index q_{ij} and the chart for the single dimension, it is possible to keep the delivering process under control, with the chance of recognizing eventual anomalies or nonphysiological disturbance causes.

It is important to highlight how the use of attribute control charts opens the door to a large quantity of literature concerning statistical product control [Grant and Leavenworth, 1988].

A procedure flowchart is shown in Figure 11.7.

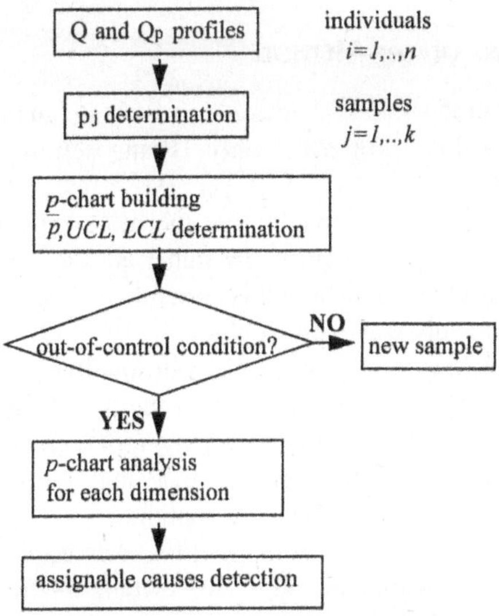

FIGURE 11.7 Procedure flowchart.

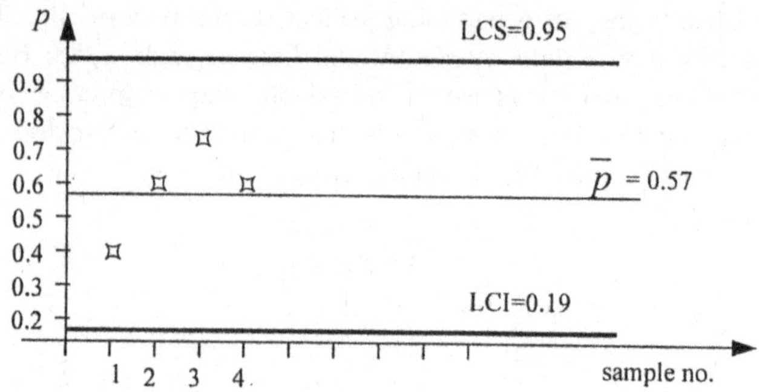

FIGURE 11.8 Control chart p for DISPEA library service.

An example may clarify this procedure. Let us assume the result from the questionnaire in the Appendix to be one among n questionnaires administered to a service customer (in this case, the DISPEA library).

At first, given the two expected $\boldsymbol{Q}_a$ = [2, 4, 4, 3, 5, 3, 2, 4] and perceived $\boldsymbol{Q}_p$ = [4, 5, 4, 5, 4, 4, 1, 5] quality profiles, the winning profile is determined by the preference system of the individual.

By applying the ELECTRE II method (see Chapter 7) and bearing in mind the importance weights for the dimensions given by the customer **W** = [3, 1, 3, 4, 5, 4, 1, 3], we found $\boldsymbol{Q}_p > \boldsymbol{Q}_a$. The q_{11} index (1° customer, 1° sample questionnaire) becomes $q_{11} = 1$.

Proceeding in this way, it is possible to determine the p_1 percentage for the first sample. By repeating the procedure for k samples by Equation (11.5), we are able to establish the limits of the control chart.

An example of a p-chart, with sample size n = 15, determined for DISPEA library service is shown in Figure 11.8. In this case, the system appears in control.

11.6.3 Implications of the Method

The example clearly shows that it is possible to evaluate an on-line control quality by means of data ordinal properties only [Franceschini and Rossetto, 1998; Fraser, 1994].

From the presentation, it is obvious that the more substantial the informative meaning given to available information, the richer are the conclusions obtained for service control. The problem lies in that this enrichment (scalarization) was artificial, so it may not be completely trustworthy.

The humbleness behavior — considering information for what they *really are*, without being *forced* to by the researcher — is the less dangerous way to analyze data, even if at first sight the result obtained may appear poor when compared with that obtained from the scalarization procedure. It may be poor, but it is closer to real knowledge of the problem of quality evolution.

The greater benefits obtainable from Qualitometro are for services involving a large number of periodic customers and in cases where the hypothesis for customer reference systems is not clear. For instance, bank services and automobile services may be included.

11.7 CONCLUSIONS

During the last few years more and more companies proffering services have endeavored to organize themselves according to models of total quality management (TQM) or of quality assurance, in a bid to develop a quality system according to the ISO series 9000 [1994] norms. Despite this effort to execute activities of planning and control of service quality, many maintain that the methodologies for quality used in the manufacturing industry can hardly be imported into the real world of services. This stems from the fact that it is very difficult to observe and monitor the quality of a service. This, however, becomes indispensable if we are to determine the specific characteristics that endow a service with the capacity to satisfy customer exigencies. This activity must be executed keeping clearly in mind a general model that explains the analogies and the differences with the world of products.

Another important concept is that of measurement: it is impossible to improve that which we are unable to describe objectively. Therefore, we must measure or evaluate all the fundamental parameters in services, because their control will help us satisfy customers.

In this context QFD is able to play a very useful role indeed. The flexibility of the technique renders it particularly suitable in the field of services. It can help to clarify the quality requested by the customer and to fix certain quality objectives through *benchmarking* on perceived quality and through technical benchmarking on offered quality.

Appendix — The Qualitometro Form

Improved service quality is the main problem for service customers. This questionnaire has two goals, evaluating what we expect and what we obtain from a service.

This questionnaire (Figure 11.9) evaluates service attributes for the DISPEA library. It is filled in two stages: before access to the service and after its delivery.

The symbols ● or □ should be written on the scale where it properly expresses expectations (with reference to an excellent library) and results for any evaluation criterion. There are eight evaluation criteria and a global service evaluation. Each criterion needs an importance weight (from 1 to 7).

An example of a questionnaire that has already been filled in is given (Figure 11.10).

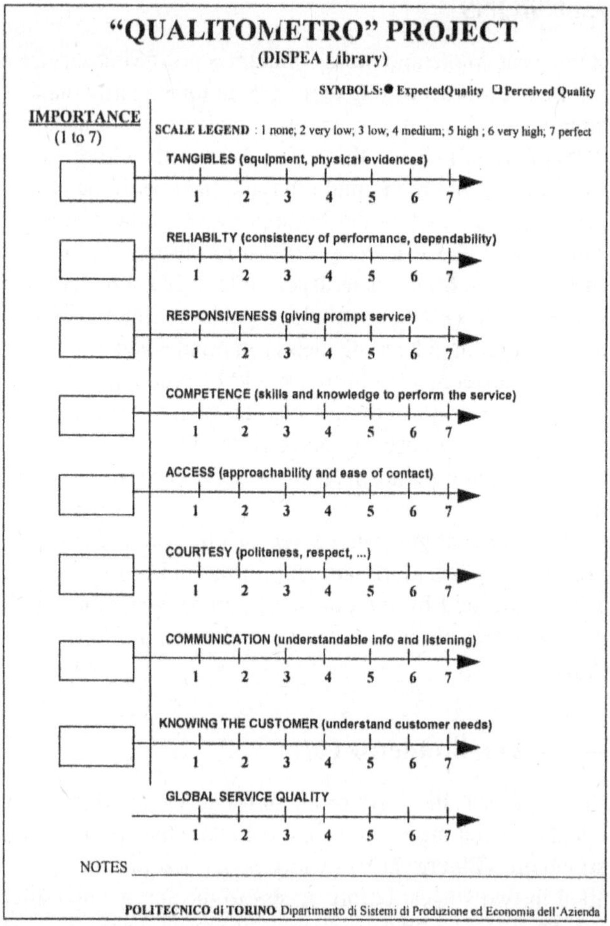

"QUALITOMETRO" PROJECT
(DISPEA Library)

SYMBOLS: ● ExpectedQuality ❑ Perceived Quality

IMPORTANCE (1 to 7)

SCALE LEGEND : 1 none; 2 very low; 3 low, 4 medium; 5 high ; 6 very high; 7 perfect

TANGIBLES (equipment, physical evidences)
1 2 3 4 5 6 7

RELIABILTY (consistency of performance, dependability)
1 2 3 4 5 6 7

RESPONSIVENESS (giving prompt service)
1 2 3 4 5 6 7

COMPETENCE (skills and knowledge to perform the service)
1 2 3 4 5 6 7

ACCESS (approachability and ease of contact)
1 2 3 4 5 6 7

COURTESY (politeness, respect, ...)
1 2 3 4 5 6 7

COMMUNICATION (understandable info and listening)
1 2 3 4 5 6 7

KNOWING THE CUSTOMER (understand customer needs)
1 2 3 4 5 6 7

GLOBAL SERVICE QUALITY
1 2 3 4 5 6 7

NOTES

POLITECNICO di TORINO Dipartimento di Sistemi di Produzione ed Economia dell'Azienda

FIGURE 11.9 Blank evaluation form for the Qualitometro method.

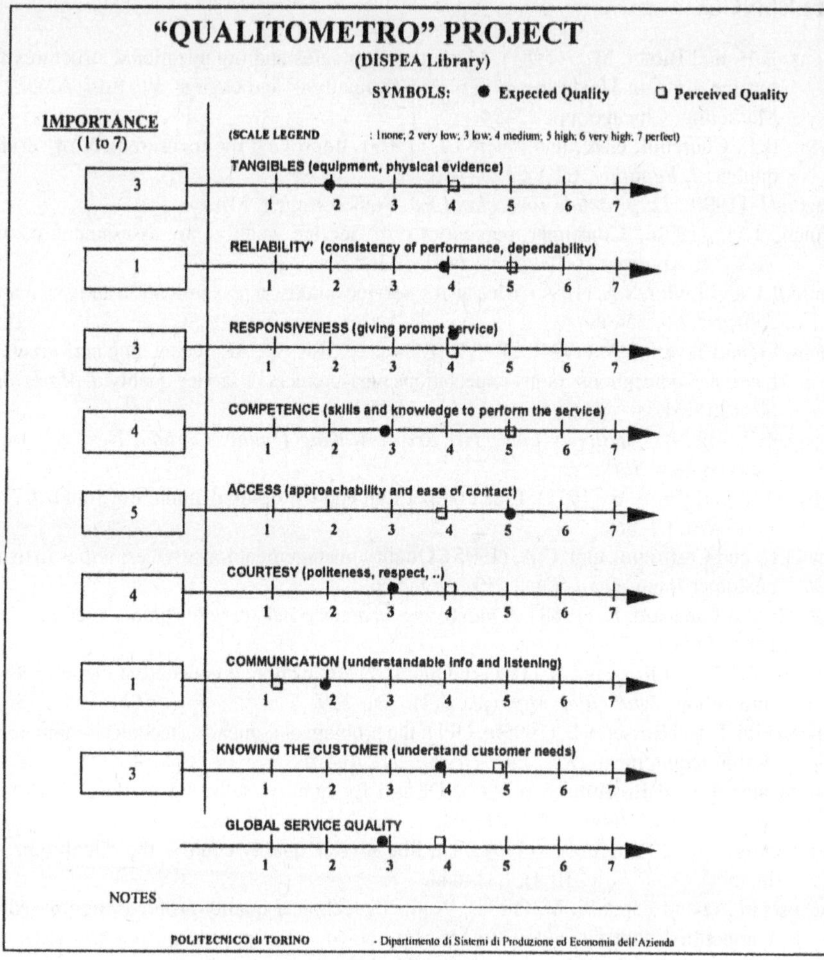

FIGURE 11.10 Example of questionnaire already filled in.

REFERENCES

Booms, B.H. and Bitner, M.J. (1981), Marketing strategies and organizational structures for service firms, in *Marketing of Services*, Donnelly, J. and George, W., Eds., American Marketing, Chicago, pp. 47–51.

Brown, B.J., Churchill, G.A., and Peter, J.J. (1993), Improving the measurement of service quality, *J. Retailing*, 69(1), 127–139.

Carlzon, J. (1990), *La piramide rovesciata*, Ed. Franco Angeli, Milano.

Carman, J.M. (1990), Consumer perceptions of service quality: an assessment of the SERVQUAL scale, *J. Retailing*, 66(1), 33–55.

Cronin, J.J. and Taylor, S.A. (1992), Measuring service quality: a reexamination and extension, *J. Mark.*, 56, 55–68.

Cronin, J.J. and Taylor, S.A. (1994), SERVPERF versus SERVQUAL: reconciling performance-based and perceptions-minus-expectations measurement of service quality, *J. Marketing*, 58, 125–131.

Crosby, P.B. (1979), *Quality Is Free: The Art of Making Quality Certain*, New American Library, New York.

Darby, M.R. and Karni, E. (1973), Free competition and the optimal amount of fraud, *J. Law Econ.*, 16, 67–86.

Drew, J.H. and Castrogiovanni, C.A. (1995), Quality management for services: issues in using customer input, *Qual. Eng.*, 7(3), 551–566.

Eigler, P. and Langeard, E. (1988), *Il marketing strategico nei servizi*, Milano, McGraw-Hill Libri Italia.

Franceschini, F. and Rossetto, S. (1995a), Quality & innovation: a conceptual model of their interaction, *Total Qual. Manage.*, 6(3), 221–229.

Franceschini, F. and Rossetto, S. (1995b), QFD: the problem of comparing technical-engineering design requirements, *Res. Eng. Design*, 7, 270–278.

Franceschini, F. and Rossetto, S. (1997), Design for quality: selecting product's technical features, *Qual. Eng.*, 9(4), 681–688.

Franceschini, F. and Rossetto, S. (1998), On-line service quality control: the 'Qualitometro' method, *Qual. Eng.*, 10(4), 633–643.

Franceschini, F. and Zappulli, M. (1998), Product's technical quality profile design based on competition analysis and customer requirements: an application to a real case, *Int. J. Qual. Reliability Manage.*, 15(4), 431–442.

Fraser, N.M. (1994), Ordinal preference representations, *Theory Decision*, 36(1), 45–67.

Garvin, D.A. (1983), Quality on the line, *Harv. Bus. Rev.*, 61, 65–73.

Garvin, D.A. (1987), Competing on the eight dimensions of quality, *Harv. Bus. Rev.*, 65(6), 101–109.

Grant, E.L. and Leavenworth, R.S. (1988), *Statistical Quality Control*, 6th ed., McGraw-Hill, New York.

Gronroos, C. (1982), *Strategic Management and Marketing in the Service Sector,* Swedish School of Economics and Business Administration, Helsingfors.

Hayes, B.E. (1992), *Measuring Customer Satisfaction*, ASQC Quality Press, Milwaukee, WI.

Kristensen, K., Kanji, G.K., and Dahlgaard, J.J. (1992), On measurement of customer satisfaction, *Total Qual. Manage.*, 3(2), 123–128.

ISO 8402 (1994), Quality Management and Quality Assurance — Vocabulary.

ISO 9000 (1994), Quality Management and Quality Assurance Standards.

ISO 9004/2 (1991), Quality Management and Quality System Elements — Guidelines for Services.

Langeard, E., Bateson, J.E.G., Lovelock, C.H., and Eiglier, P. (1981), *Service Marketing: New Insights from Consumers and Managers,* Marketing Science Institute, Cambridge, MA.

Larichev, O.I., Moshkovich, H.M., Mechitov, A.J., and Olson, D.L. (1993), Experiments comparing qualitative approaches to rank ordering of multiattribute alternatives, *J. Multi-Criteria Decision Anal.,* 2(1), 5–26.

Larichev, O.I., Olson, D.L., Moshkovich, H.M., and Mechitov, A.J., (1995), Numerical vs. cardinal measurements in multiattribute decision making: how exact is enough, *Organ. Behav. Hum. Decision Proc.,* 64(1), 9–21.

Lehtinen, U. and Lehtinen, J.R. (1982), Service Quality: A Study of Quality Dimensions, working paper unpublished, Service Management Institute, Helsinki.

Levitt, T. (1992), *Marketing Imagination*, Sperling & Kupfer, Milano, pp. 62–92,.

Lewis, R.C. and Booms, B.H. (1983), The marketing aspects of service quality, in *Emerging Perspectives on Service Marketing*, Berry, L., Shostack, G., and Upah, G., Eds., American Marketing, Chicago, pp. 99–107.

McConnell, J.D. (1968), Effect of pricing on perception of product quality, *J. Appl. Psychol.*, 52, 300–303.

Negri, L. (1992), La qualità nelle imprese di servizi, in *Spazio Impresa*, 26, 20–24.

Nelson, P. (1974), Advertising as information, *J. Political Econ.*, 81, 729–754.

Oliver, R. (1981), Measurement and evaluation of satisfaction process in retail settings, *J. Retailing*, 57(1), 25–48.

Olander, F. (1970), The influence of price on the consumer's evaluation of products, in *Pricing Strategy,* Taylor, B. and Willis, G., Eds., Brandon-Systems Press, Princeton, NJ.

Ostanello, A. (1985), Outranking methods, in *Multiple Criteria Decision Methods and Application*, Fandel, G. and Spronk, J., Eds., Springer-Verlag, Berlin, pp. 41–60.

Parasuraman, A., Zeithaml, V.A., and Berry, L.L. (1985), A conceptual model of service quality and its implications for future research, *J. Mark.*, 49, 41–50.

Parasuraman, A., Zeithaml, V.A., and Berry, L.L. (1988), SERVQUAL: a multiple-item scale for measuring consumer perception of service quality, *J. Retailing*, 64(1), 12–40.

Parasuraman, A., Berry, L.L., and Zeithaml, V.A. (1991), Refinement and reassessment of the SERVQUAL scale, *J. Retailing*, 67(4), 420–450.

Parasuraman, A., Berry, L.L., and Zeithaml, V.A. (1993), More on improving service quality measurement, *J. Retailing*, 69(1), 140–147.

Parasuraman, A., Zeithaml, V.A., and Berry, L.L. (1994), Reassessment of expectations as a comparison standard in measuring service quality: implications for future research, *J. Mark.*, 58(1), 11–124.

Parasuraman, A., Berry, L.L., and Zeithaml, V.A. (1996), The behavioral consequences of service quality, *J. Mark.*, 60, 31–46.

Peters, T.J. and Waterman, R.H., Jr. (1984), *Alla ricerca dell'eccellenza — Lezioni dalle aziende meglio gestite*, Sperling & Kupfer, Milano.

Roy, B. (1991), The outranking approach and the foundations of ELECTRE methods, *Theory Decision,* 31(1), 49–73.

Rupil, A. (1996), Metodi e tecniche per la misurazione della qualità degli attributi di un prodotto: un'applicazione nel settore dei veicoli industriali, thesis degree, Politecnico di Torino.

Sasser, W.E. Jr., Olsen, P., and Wyckoff, D.D. (1978), *Management of Service Operations: Text and Cases,* Allyn & Bacon, Boston.

Schvaneveldt, S.J., Enkawa, T., and Miyakawa, M. (1991), Consumer evaluation perspectives of service quality: evaluation factors and two-way model of quality, *Total Qual. Manage.*, 2(2), 149–161.

Teas, R.K. (1993), Expectations, performance, evaluation, and consumers' perceptions of quality, *J. Mark.*, 57, 55–68.

Teas, R.K. (1994), Expectations as a comparison standard in measurement of service quality: an assessment of a reassessment, *J. Mark.*, 58, 132–139.

Terzago, M. (1995), Quality Function Deployment: da strumento organizzativo a strumento di supporto alla progettazione, thesis degree, Politecnico di Torino.

Zeithaml, V.A. (1981), How consumer evaluation process differ between goods and services, in *Marketing of Services*, Donnelly, J. and George, W., Eds., American Marketing, Chicago, pp. 186–190.

12 Application of Quality Function Deployment to Industrial Training Courses

12.1 INTRODUCTION

In the industrial world, *training* refers to the whole group of intentional actions — in the classroom or during the actual work (*training on the job*) — to increase knowledge and *individual* capabilities of human resources [Braga and Roncari, 1994].

The spreading of training programs based on the total quality management (TQM) approach, particularly significant between the late 1980 s and the early 1990s, has given a meaningful contribution to *the way to train* in the industrial world. In fact, in this period of economic slump, companies have felt the necessity to invest in changes that could offer them real competitive advantages for the future. Training has become, even more than before, a *strategy for optimizing companies' outputs according to market requests*.

The demand of industrial training courses is growing more qualified, whereas requests for the realization of training programs in the field are becoming even more definite, and having precise improvement goals consistent with each company's quality plan.

In addition to these requirements, the customer has shown expectations of measurable outcomes and of return on investment that are more relevant now in spite of other past training programs, which were less "aimed."

At present, some strategic changes are in progress in the industrial training world, such as shortening learning cycle, developing a learning environment (the so-called *learning organization*), improving the flexibility of companies' reply to the market, developing decentralization and industrialization of training programs, and finally improving evaluation systems.

12.2 DIFFERENT CUSTOMERS WITH DIFFERENT NEEDS

In our specific case, quality function deployment (QFD) has been used for the design of a *theoretical–practical* training course, with the purpose of giving a well-defined knowledge target to all the participants [Franceschini and Terzago, 1998].

To better clarify the problem, we may think about the design of a training course on statistical process control (SPC), covering a period of a week. This course may include, for example, some theoretical topics of basic statistics, with the aim of getting practical capabilities through the creation of data collection sheets, bar charts, filled in control charts for variables or attributes, and so on.

As we have remarked on previous chapters, the very first step of QFD requires the determination of customer needs. Thus, it is necessary to state the groups of customers that we address, with the purpose of satisfying their needs.

From this point of view, we immediately notice that the world of industrial training courses is very diversified, because of the presence of many "actors" taking part in the process. They are:

- The *participants* (or *customers–users*), who are the persons taking part in the training course for different reasons: subsequent to their company's will, for professional upgrade, for personal interest, and so on
- The *investor* (or *customer–investor*), which is the company that *decides* to pay a training agency for a training course and that, in case of a tailor-made training course, agrees on the basic subjects with the expert
- The *training agency* (or *service provider*), which is the service company that organizes the training course
- The *expert* (or *technician*), which is the person (an external consultant or an internal employee of the training agency) who is skilled in the specific subject of the course and who designs its detailed contents
- The *teacher*, who is the professional chosen by the training agency — often according to the specifications defined by the expert — to provide all or part of the training service to the participants
- The *recruiting manager* (or *customer selector*), usually belonging to the human resources department, whose job is to select the group of participants to the training course

All these actors play different roles and have, with respect to a training course, different kind of requirements.

For example, the participants may want:

- To improve their professional know-how and capabilities
- To learn new things, even if never heard before
- To check and measure their knowledge
- To be entertained in a pleasant way

The investor wants, for example, the return of investment, in terms of:

- Added value in the process activities
- Know-how and new professional capabilities
- New expertise in solving specific problems

Furthermore, the investor wants, for example, to entail participants to improve:

- Involvement
- Company membership
- Sense of belonging to a group or a professional community

On the other hand, the investor appreciates these service characteristics:

- Clearness and easy understanding of the offer (especially in the case of a course sold by catalog)
- Fairness of selling price
- Timeliness in providing
- Ease of access
- Precise definition of the training targets
- Others

The expert wants to insert the arguments considered most important to design the body of the course. For example the expert appreciates:

- Targets definition
- Information about the participants
- Basic timing for the realization of the course
- Information about availability of technical and logistical instruments
- Information about teachers' profiles

On the other hand, the teacher wants to be the "main actor" in that particular course situation, with those contents and with that specific teaching methodology. The teacher enjoys:

- Ease of providing (customization of the service based on the teacher, for example, course not structured for a professional, well-engineered for a nonexpert)
- Complete information about the contents of the course
- Previous knowledge of all participants
- Information about rules and style of the investor and the training agency
- Previous knowledge of other teachers taking part in the training course
- Complete information about other modules or subjects linked with own subject
- Complete information about the investor
- Acknowledgment and sharing of investor's purposes
- Chance for growing professionally
- Complete documentation as support
- Fair payment
- Ease in service planning
- Good relationship with provider's processes
- Clearness
- Assistance by the training agency
- Perfect knowledge of the teaching message to be sent
- Conformity of service to own accepted mental scheme

Because of all these different actors involved, each having special needs to satisfy, we decided to identify the participant (the worker) as the *focus* (the area of intervention) of our work, with the performances and the know-how of the single human resource as the real *target* of a training course.

12.3 CUSTOMER SATISFACTION ANALYSIS

We used many sources to define what could be the requests of participants (customers–users) for a standard training course, from well-known evaluation forms or questionnaires to be filled in at the end of a course, to previous analysis of *customer satisfaction* (CS) made by different training agencies [Baiardi, 1994].

This analysis yielded the creation of a list of 49 *desiderata* (desires) [Franceschini and Terzago, 1998] expressed by a participant of a training course. These data were considered as raw data [Akao, 1990], from which we started to recognize customers' needs. To do that, we took great care not to confuse these requirements with *proposals of solution* or specifications for designing a service suggested by the customers.

In fact, individual participants of a training course may say, for example, they like to be given a certificate of final evaluation, or an attendance certificate, or even to hear a linear description instead of one with many logical leaps. Nevertheless, these elements do not represent customers' needs, but instead some characteristics for providing the service, or some specifications that are related to other implicit needs — such as the utility of the training course, the ease of certification of the acquired know-how or capabilities, or the ease in following the teacher's speech.

12.4 DEMANDED QUALITY CHART

At this point, by similarity we grouped sets of all the requirements we obtained from the translation of raw data into reworded data, using an affinity diagram [Urban and Hauser, 1993].

In this way, we grouped all the lower (third) level requirements (*detailed needs*) into higher level categories (tactical or *second-level needs*) and then into strategic (or *primary*) needs. The labels of the various categories obtained were chosen, when possible, using the names of requirements already present in the list; otherwise we decided to define a more suitable label.

Strategic (or primary) needs that we pointed out (according to our customer satisfaction analysis) for a standard participant of a training course are:

1. *To be useful* (i.e., effective, fit to use)
2. *To deal with interesting subjects* (causing participants' interest and curiosity)
3. *To be pertinent and consistent* with customer expectations
4. *To be clear* (concept dealing with the methods for the transmission of the training message, and referring to the ability of finding out the meanings of the terms used in the explanation)
5. *To be comprehensible* (concept dealing with the learning process and referring to the ability to fix a relationship between explained ideas and other ideas already present in the minds of the participants; in fact the word coming from the Latin *comprehendo*, which means "to embrace" with one's mind)

6. *To be well organized*
7. *To be well managed* (referring to the process of providing the service in the classroom and to relationships between teacher and participants and among participants)
8. *To entertain in a pleasant way* (aspects referring to the emotional point of view connected with the participation to the course; not optional elements, but very important because they constitute for the participant an emotional inclination to get all the know-how and capabilities that the course aims to provide)

All these eight macrocategories are made up of second-level requirements and possibly of third-level ones (used to better describe some second-level requirements). It is often difficult to determine which category is proper to a specific requirement, because some needs intersect two or more categories; moreover, the same categories are not completely independent. One should not be discouraged by this difficulty; in fact, the most important aspect is not the construction of a perfect hierarchy of customer requirements, but it is not forgetting any of the elements issued from the CS analysis.

Table 12.1 shows the demanded quality chart for an industrial training course.

12.5 SERVICE CHARACTERISTICS CHART

To define technical characteristics we had to use to design a new training course, which would be able to satisfy all requirements we mentioned before, we proceeded as indicated in Chapter 4.

We took into consideration all the third-level requirements and the second-level ones that had no lower specification; and for each, we pointed out the characteristics of the training course needed to assure customer satisfaction (Table 12.2).

Afterward, we grouped together all the characteristics previously defined in similar categories, using three levels of detail, with the same technique described for the creation of the demanded quality chart.

This yielded eight macrocategories or warning areas that we used to start the design process of the training service. These are:

1. *Contents* (necessary to recognize them in a proper way, to verify them, and finally to put them into the course body structure)
2. *Supports* (including both training books and technical supports for teaching)
3. *Providing* (in the phases of explanation, examples, exercises, industrial case studies, and tests)
4. *Learning* (critical contents, logical steps, etc.)
5. *Class management* (relationship with the class, keeping its attention, working-time management, etc.)
6. *Communication* (language and style used in the explanation)
7. *Organization* (course duration, timetable, composition of participants in groups, certificates, additional services, evaluation questionnaires, etc.)
8. *Logistics* (classroom functionality and comfort, easily reached site, etc.)

TABLE 12.1
Demanded Quality Chart for an Industrial Quality Service

1st Level Requirements	2nd Level Requirements	3rd Level Requirements	Kano	Weight	AHP
1) To be useful	1.1) Providing useful knowledge	1.1.1) Providing useful theoretical concepts	O	4	0.014
		1.1.2) Providing useful technical/practical capabilities	B	5	0.021
	1.2) Effective in solving the training problem	1.2.1) Gives precise answers to concrete company problems	B	5	0.025
		1.2.2) Is transmissible/ applicable to the participant's work life	B	5	0.039
		1.2.3) Allows to recognize oneself in the situations described	E	3	0.008
	1.3) Useful for the certification of acquired know-how and capabilities	1.3.1) Produces a formal acknowledgment	E	3	0.013
		1.3.2) Produces an "easy-to-spend" training result	O	4	0.013
	1.4) Allows to verify and measure one's own knowledge	1.4.1) Allows to perceive the progress in the learning process	O	4	0.006
		1.4.2) Allows to self-evaluate one's learning	O	4	0.006
2) To deal with interesting subjects	2.1) Allows to examine interesting subjects in deep	2.1) Allows to examine interesting subjects in deep	O	4	0.054
	2.2) Is full of new things	2.2.1) Gives more new information	O	4	0.008
		2.2.2) Allows learning "to do new things"	E	3	0.015
	2.3) Is stimulating	2.3.1) Provides an easy, but not "commonplace" learning	O	4	0.015
		2.3.2) Stimulates new ideas	E	3	0.016
		2.3.3) Stimulates new training requirements	E	3	0.003
	2.4) Allows the comparison with different environments	2.4.1) Makes possible a comparison with other realities through the teacher	O	4	0.004

TABLE 12.1 (continued)
Demanded Quality Chart for an Industrial Quality Service

1st Level Requirements	2nd Level Requirements	3rd Level Requirements	Kano	Weight	AHP
		2.4.2) Makes possible a comparison with other realities through the participants	O	4	0.004
		2.4.3) Makes possible a comparison with other realities through an industrial stage	E	3	0.022
3) To be pertinent/consistent	3.1) The course contents are "organized"	3.1) The course contents are "organized"	B	5	0.057
	3.2) Course program consistent with participants expectations	3.2) Course program consistent with participants expectations	B	5	0.086
4) To be clear	4.1) Clearness of the used terminology	4.1) Clearness of the used terminology	B	5	0.018
	4.2) Considering educational differences among the participants	4.2) Considering educational differences among the participants	B	5	0.027
5) To be comprehensible	5.1) Teacher easy to follow in his reasoning	5.1.1) Explanation easy to follow	B	5	0.014
		5.1.2) Explanation not taking unknown subjects for granted	B	5	0.014
		5.1.3) Explanation "aimed to" (it is clear where we want to go and through which way)	O	4	0.014
	5.2) Easy to comprehend the subject one is dealing with	5.2) Easy to comprehend the subject one is dealing with	O	4	0.027
	5.3) Logical path easy to step back	5.3) Logical path easy to step back	O	4	0.03
6) To be well organized	6.1) "Comfortable" timetable	6.1) "Comfortable" timetable	O	4	0.038
	6.2) Putting at one's disposal up to date technical instruments	6.2) Putting at one's disposal up to date technical instruments	O	4	0.023
	6.3) Classroom fit to use	6.3.1) Set in a comfortable classroom	B	5	0.01
		6.3.2) Set in a practical classroom	B	5	0.01
	6.4) Documentation fit to use	6.4.1) Complete teaching documentation	O	4	0.012

TABLE 12.1 (continued)
Demanded Quality Chart for an Industrial Quality Service

1st Level Requirements	2nd Level Requirements	3rd Level Requirements	Kano	Weight	AHP
		6.4.2) Teaching documentation clear/readable	O	4	0.012
		6.4.3) Teaching documentation easy to consult	O	4	0.012
	6.5) Provides complete optional services	6.5) Provides complete optional services	E	3	0.009
	6.6) Site easy to reach	6.6) Site easy to reach	O	4	0.01
	6.7) Possibility to express one's own evaluation upon course quality	6.7) Possibility to express one's own evaluation upon course quality	O	4	0.006
7) To be well managed	7.1) Well-balanced work timetable	7.1) Well-balanced work timetable	O	4	0.005
	7.2) Dedicates enough time to every subject of the program	7.2) Dedicates enough time to every subject of the program	O	4	0.018
	7.3) Allows everyone to express oneself	7.3) Allows everyone to express oneself	B	5	0.04
	7.4) Teacher is available	7.4.1) Teacher available to provide help	O	4	0.013
		7.4.2) Teacher available to answer to questions made by the participants	B	5	0.01
		7.4.3) Teacher available to let the participants express themselves	O	4	0.008
	7.5) "Pleasant" relationship with the class	7.5) "Pleasant" relationship with the class	O	4	0.027
	7.6) "Leadership" of the teacher	7.6) "Leadership" of the teacher	O	4	0.023
8) To entertain in a pleasant way	8.1) Nice teacher	8.1) Nice teacher	E	3	0.024
	8.2) Nice group of participants	8.2) Nice group of participants	E	3	0.034
	8.3) Pleasant environment	8.3) Pleasant environment	E	3	0.017
	8.4) Allows acknowledgment from the group of participants	8.4) Allows acknowledgment from the group of participants	O	4	0.027
	8.5) Stimulates involvement	8.5) Stimulates involvement	O	4	0.041

TABLE 12.2
Extracting Service Characteristics from the Analysis of Customer Requirements

Customer Requirements	Service Characteristics
Providing useful theoretical concepts	Identity of all theoretical concepts that are at the basis of the general contents of the program
	Inclusion of the theoretical concepts previously recognized into the course program
	Recall of useful theoretical concepts during the explanation phase
...	...
Ease of explanation	Use of slides as support
	Delivery time of textbooks to the participants
	Type of textbooks
	Clearness of the explanation
	Level of detail of the explanation
	Linearity of the explanation
	Scanning speed of logical steps

In Table 12.3 we show, for example, the house of quality (HoQ) for the macroarea supports.

12.6 PRIORITIZATION OF SERVICE QUALITY CHARACTERISTICS

According to our analysis concerning evaluation questionnaires for the quality of training courses, we proved that the global satisfaction of participants cannot be expressed as a linear combination of the degree of satisfaction of each requirement (explicit or implicit) that they may feel concerning the service. Customers are not usually able to give cardinal evaluations of the relative importance of requirements. Furthermore, as we have seen in Chapters 7 and 11, aggregating the scores given by different customers may be quite a difficult job [Franceschini and Rossetto, 1995a, 1997, 1998].

In our specific case, according to Kano's model [Kano et al., 1984], we classified all requirements into three categories: B, O, and E (see Chapter 4, Section 4.5.1). Once this classification is made, the service provider can decide whether to favor basic (B) or excitement (E) customer requirements. In our case, for example, we chose to privilege the satisfaction of B requirements, giving them a score of five; O requirements, a score of four; and E requirements, a score of three. The assigned scores have been put in the "weight" column in Table 12.1.

A different approach to the prioritization involved determining the weight values with the use of the analytical hierarchy process (AHP) method (see Chapter 5). This solution is very interesting in situations with small numbers of decision makers. Thus, it could be applied in the case of a training course provided on order, and not in the case of a training course sold by catalog. The weights we obtained applying the AHP method to the resulting comparison matrix are shown in the "AHP" column in Table 12.1.

TABLE 12.3
House of Quality for the Supports Area

1st level Characteristics	2) Supports									
2nd level Characteristics	2.1) Supports for the Providing Phase					2.2) Teaching Documentation				
3rd level Characteristics / 3rd level Requirements	2.1.1) Slides	2.1.2) Teacher's Manual	2.1.3) Participant's Manual	2.1.4) Video cassette	2.1.5) Educational software	2.2.1) Quantity of the teaching documentation	2.2.2) Delivery time of the documentation	2.2.3) Type of documentation	2.2.4) Reference bibliography	2.2.5) Preparing supplementary documentation
1.1.1) Providing useful theoretical concepts										
1.1.2) Providing useful technical/practical capabilities					○					
1.2.1) Gives precise answers to concrete company problems										
1.2.2) Is transmissible/applicable to the participant's work life										
1.2.3) Allows to recognize oneself in the situation described										
1.3.1) Produces a formal acknowledgment										
1.3.2) Produces an "easy-to-spend" training result										
1.4.1) Allows to perceive the progress in the learning process										
1.4.2) Allows to self-evaluate one's learning										

2.1) Allows to examine interesting subjects in deep									●	●
2.2.1) Gives more new information										
2.2.2) Allows learning "to do new things"										
2.3.1) Provides an easy, but not "commonplace" learning	●			●						
2.3.2) Stimulates new ideas										
2.3.3) Stimulates new training requirements										
2.4.1) Makes possible a comparison with other realities through the teacher										
2.4.2) Makes possible a comparison with other realities through the participants										
2.4.3) Makes possible a comparison with other realities through an industrial stage										
3.1) The course contents are "organized"										
3.2) Course program consistent with participant's expectations										
4.1) Clearness of the used terminology										
4.2) Considering educational differences among the participants										
5.1.1) Explanation easy to follow	●	●		○	○		●	●		
5.1.2) Explanation not taking unknown subject for granted										
5.1.6) Explanation "aimed to" (it is clear where we want to go and through which way)		●								
5.2) Easy to comprehend the subject one is dealing with	●						○			
5.3) Logical path easy to step back	△		●							
6.1) "Comfortable" timetable										
6.2) Putting at one's disposal up to date technical instruments				●	●					

TABLE 12.3 (continued)
House of Quality for the Supports Area

1st level Characteristics	2) Supports									
2nd level Characteristics	2.1) Supports for the Providing Phase					2.2) Teaching Documentation				
3rd level Characteristics / 3rd level Requirements	2.1.1) Slides	2.1.2) Teacher's Manual	2.1.3) Participant's Manual	2.1.4) Video cassette	2.1.5) Educational software	2.2.1) Quantity of the teaching documentation	2.2.2) Delivery time of the documentation	2.2.3) Type of documentation	2.2.4) Reference bibliography	2.2.5) Preparing supplementary documentation
6.3.1) Set in a comfortable classroom										
6.3.2) Set in a practical classroom										
6.4.1) Complete teaching documentation			●			●			○	●
6.4.2) Teaching documentation clear/readable			○			●		●		
6.4.3) Teaching documentation easy to consult			●			●	●			
6.5) Provides complete optional services										
6.6) Site easy to reach										
6.7) Possibility to express one's own evaluation upon course quality										
7.1) Well-balanced work timetable										
7.2) Dedicates enough time to every subject of the program	○	●								
7.3) Allows everyone to express oneself										

7.4.1) Teacher available to provide help										
7.4.2) Teacher available to answer questions asked by the participants										
7.4.3) Teacher available to let the participants express themselves										
7.5) "Pleasant" relationship with the class										
7.6) "Leadership" of the teacher										
8.1) Nice teacher										
8.2) Nice group of participants										
8.3) Pleasant environment										
8.4) Allows acknowledgment from the group of participants										
8.5) Stimulates involvement				△	△					
Rating according to "Weight"	133	117	120	91	70	108	93	81	48	72
Relative Importance	1.96%	1.72%	1.77%	1.34%	1.03%	1.59%	1.37%	1.19%	0.71%	1.06%
(Ranking according to "Weight")	[1]	[3]	[2]	[6]	[9]	[4]	[5]	[7]	[10]	[8]
Rating according to "AHP"	0.58383	0.40569	0.52095	0.42104	0.35065	0.32261	0.31318	0.23144	0.52384	0.59554
(Raking according to "AHP")	[2]	[6]	[4]	[5]	[7]	[8]	[9]	[10]	[3]	[1]

Legend: Relationship matrix between customer requirements and service characteristics/service design activities

● Strong relationship = Score 9
○ Medium relationship = Score 3
△ Weak relationship = Score 1

12.7 SOME RESULTS

To show the relationships between customer requirements and service characteristics, we used the typical QFD symbols presented in Chapter 4. By filling the relationship matrix, we found the job quite heavy because this matrix has 50 × 109 = 5,450 cells.

The prioritization of service characteristics has been made with the traditional method (or *independent scoring method*) (see Chapter 4) using the 1 to 3 to 9 scale. By comparing the two types of prioritization, obtained using weights found with the AHP method (B prioritization) and the Kano model (A prioritization), respectively, we can see quite similar rankings for each of the eight macroareas identified.

In the *contents* macroarea, the most important design activity is pointing out all the theoretical concepts that are at the basis of the general contents of the program (according to A prioritization), whereas according to B prioritization pointing out the teaching goals of the course (ranked second in A prioritization) is most important.

In the *supports* macroarea, the use of slides (ranked first in A prioritization and second for B prioritization) has very great importance. In the *providing* area the most important design items are the collection of participants' expectations during the presentation of the course, the concrete form of the explanation, and the logical connection of the examples with the specific professional background of participants.

In the *learning* macroarea the scanning speed of logical steps (ranked first both in A and B prioritization) and the presence of process checks are the most important characteristics, whereas in the *class management* macroarea the space to investigate interest matters and the teaching methods rotation have great importance.

Teachers' behavior and using known words are the most important characteristics in the *communication* area, whereas in the *organization* area the presence of a qualification certificate after attending the course is the most favored characteristic. Particularly, in the organization area, A prioritization is quite similar to B prioritization.

In considering the *logistics* macroarea of the course, the most important characteristic is the comfort (brightness, noises, and warmth) of the classroom.

12.8 FINAL CONSIDERATIONS ABOUT THE CASE STUDY

Thinking about customer requirements and their conversion to design characteristics may contribute, in a very efficient way, to making the attributes of an industrial training course much more definite.

Moreover, our experiment, which consists of the application of QFD to a service, permits us to note two large differences with respect to a traditional QFD employment for a product development:

1. Confusion exists among elements of technical quality (concerning what the customer really gets from the service) and functional quality (concerning the way in which the service is provided) [Gronroos, 1982].
2. Difficulty exists in measuring the level of many service characteristics. In fact, in contrast to what happens for product design, only some of the technical characteristics we determined (for example comfort (brightness,

noises and heat) of the classroom) are easily measurable and can be standardized, whereas we can hardly do the same for the majority of other characteristics. Moreover, some characteristics are measurable just as users' perceptions (i.e., nice environment and teachers' behavior).

From the case study we can see that for a training course the final outcome also depends on customer behavior during the providing phase. For this reason a training agency cannot often ensure a good result, but it can assure that everything necessary to reach the expected result will be done. The design process becomes, therefore, a fundamental step, in which it is possible to carry out a prevention activity.

Among the limits we can find in our specific application, there is surely the problem of having the relationship matrix too large. As first approximation, we have not divided into more steps the input–output process of translating customer requirements into design specifications. One possible solution to make the process leaner could be to divide it into three phases: design of contents, design of providing process, and design of quality control process (Figure 12.1).

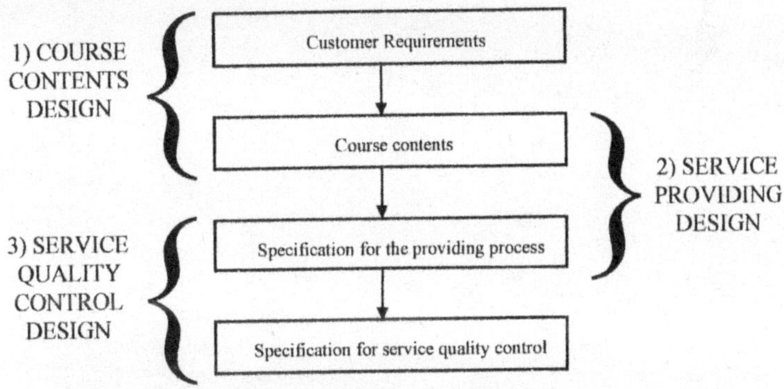

FIGURE 12.1 Steps for designing a training course using QFD.

REFERENCES

Akao, Y. (1990), *Quality Function Deployment,* Productivity Press, Cambridge, MA.

Baiardi, D. (1994), *Guida allo Sviluppo di un Sistema Qualità per le Organizzazioni di Servizi Formativi*, Ed. Unione Industriale di Torino.

Braga, G. and Roncari, A. (1994), Formazione e qualità, *De Qualitate*, 3(7), 23–30.

Franceschini, F. (1998), *Quality Function Deployment: uno strumento concettuale per coniugare qualità e innovazione*, Il Sole 24 ORE Libri, Ed., Milano.

Franceschini, F. and Rossetto, S. (1995a), QFD: the problem of comparing technical-engineering design requirements, *Res. Eng. Design*, 7, 270–278.

Franceschini, F. and Rossetto, S. (1995b), Quality & innovation: a conceptual model of their interaction, *Total Qual. Manage.*, 6(3), 221–229.

Franceschini, F. and Rossetto, S. (1997), Design for quality: selecting product's technical features, *Qual. Eng.*, 9(4), 681–688.

Franceschini, F. and Rossetto, S. (1998), On-line service quality control: the qualitometro method, *Qual. Eng.*, 10(4), 633–643.

Franceschini, F. and Terzago, M. (1998), An application of quality function deployment to industrial training courses, *Int. J. Qual. Reliability Manage.*, 15(7), 753–768.

Gronroos, C. (1982), Strategic Management and Marketing in the Service Sector, Swedish School of Economics and Business Administration, Helsingfors.

Kano, N., Seraku, N., Takahashi, F., and Tsuji, S. (1984), Attractive quality and must-be quality, *J. Jpn. Soc. Qual. Control,* 14(2), 39–48.

Saaty, T.L. (1990), *Decision Making for Leaders: The Analytic Hierarchy Process for Decisions in a Complex World,* (rev. ed.), RWS Publications, Pittsburgh.

Urban, G.L. and Hauser, J.R. (1993), *Design and Marketing of New Products,* Prentice Hall, Englewood Cliffs, NJ.

Index

D

Q

R

S